Misbehaving Science

Misbehaving Science

*Controversy and the Development
of Behavior Genetics*

AARON PANOFSKY

THE UNIVERSITY OF CHICAGO PRESS   CHICAGO AND LONDON

AARON PANOFSKY is assistant professor in the Department of Public Policy and Institute for Society and Genetics at the University of California, Los Angeles.

The University of Chicago Press, Chicago 60637
The University of Chicago Press, Ltd., London
© 2014 by The University of Chicago
All rights reserved. Published 2014.
Printed in the United States of America
23 22 21 20 19 18 17 16 15 14   1 2 3 4 5

ISBN-13: 978-0-226-05831-3 (cloth)
ISBN-13: 978-0-226-05845-0 (paper)
ISBN-13: 978-0-226-05859-7 (e-book)
DOI: 10.7208/chicago/9780226058597.001.0001

Library of Congress Cataloging-in-Publication Data

Panofsky, Aaron, author.
  Misbehaving science : controversy and the development of behavior genetics / Aaron Panofsky.
    pages cm
  Includes bibliographical references and index.
  ISBN 978-0-226-05831-3 (cloth : alk. paper) — ISBN 978-0-226-05845-0 (pbk. : alk. paper) — ISBN 978-0-226-05859-7 (e-book)   1. Behavior genetics—Social aspects.
2. Geneticists—Social aspects.   I. Title.
  QH457.P36 2014
  576.5—dc23
                                                            2013034372

Contents

Illustrations

Acknowledgments

For this project, long in the making, I have gathered many debts. My thanks go out first to the many behavior geneticists, critics, and commentators who spoke with me generously and openly about the field in both formal interviews and less formal conversations at conferences and workshops. Their words were the most important source of data for this study. I also appreciate invitations to participate in and observe several meetings of the "Crafting Tools for Public Conversation about Behavioral Genetics" project of the Hastings Center. The Rockefeller Archive Center, the American Philosophical Society, and the online Profiles in Science collection at the National Library of Medicine provided archival materials revealing important parts of behavior genetics' history. Data about grants in behavior genetics came from the now defunct online open government project, the Sunshine Project. For generous funding I must thank the National Science Foundation, the Graduate School of Arts and Sciences and the Institute for the History of the Production of Knowledge at New York University, the Robert Wood Johnson Foundation, and the UCLA Faculty Career Development Fund.

I have been blessed with a fantastic and dedicated set of mentors over the years. At NYU Richard Sennett, David Garland, Edward Lehman, Rayna Rapp, Craig Calhoun, and Troy Duster encouraged but also challenged me in ways that sharpened my thinking. Troy and Craig I thank especially for their confidence in me, their patience, and the many times they helped steer me back on course when I was heading off into the wilderness. This project was inspired by my conversations with Dorothy Nelkin long ago. She gave me a good push at the beginning but passed away long before the project was completed. I'm sure the book would have been better and been finished quicker had Dot been there to keep helping me along; I'm still sad not to be able to share the results with her.

I've had great colleagues to work with over the years. Thanks to Courtney Abrams, Michael Armato, Neal Caren, Amie Hess, Monika Krause, and Michael McQuarrie for their reads and critiques of the project early on. It has also been an honor to be a part of and present my work to the NYLON Culture Research Network led by Craig Calhoun and Richard Sennett. When I was a fellow at UC Berkeley, I benefited from conversations with sociology colleagues John Levi Martin, Michael Burawoy, Marion Fourcade, Dan Dohan, Anthony Chen, and Cybelle Fox as well as my political science, economics, and policy colleagues Rob Mickey, Naomi Murakawa, Michael Anderson, David Kirp, and John Elwood.

At UCLA I've had the great fortune to be a part of two stimulating, supportive, and collegial units. I'd like to thank my colleagues in Public Policy for their steady encouragement even though my work differs so much from theirs. Conversations with many of them, but in particular Mark Peterson, Andy Sabl, Mark Kleiman, and Al Carnesale, helped me think about my work in ways that could highlight its policy relevance. Colleagues at the Institute for Society and Genetics have been unwavering in their faith and interest in the project even during long periods when it didn't look like it was progressing much. The book would never have appeared in this form without the ideas, advice, and questions of Patrick Allard, Soraya DeChadaravian, Christopher Kelty, Hannah Landecker, Jessica Lynch-Alfaro, Ed McCabe, Linda McCabe, John Novembre, Christina Palmer, Abigail Saguy, David Schleifer, Norton Wise, and Stefan Timmermans. Conversations with Gil Eyal, John Dupré, Stefan Helmreich, David Moore, Kelly Moore, Alondra Nelson, Nicole Nelson, Diane Paul, Sarah Richardson, Nikolas Rose, and Karen-Sue Taussig were very helpful. I've also presented pieces of the project to audiences at NYU, the New School, the Hastings Center, London School of Economics, UCLA, York University (Toronto), UC Berkeley, National University of Singapore, and various American Sociological Association and Society for Social Studies of Science conferences; I am most grateful for their attention, questions, and suggestions.

There are a few colleagues whom I'd like to offer special thanks for shaping my ideas in conversations over the years and for reading drafts of my work. I was privileged to be part of an incredibly dedicated and productive writing group with Ruha Benjamin, Catherine Bliss, and

Sara Shostak. They kept me moving forward and focused on the arguments that matter during the hard slog of turning out the manuscript. It's hard to overestimate the impact over the years of Michael McQuarrie and Monika Krause on my ideas; they're like extra lobes of my brain. And Hannah Landecker at a crucial moment basically saved the project: she helped me see and understand the deficits of the first draft and to restructure the entire narrative into its current historical and controversy-focused organization. And then she commented on drafts of most of the chapters, some more than once, thus ensuring I wouldn't botch the job.

It has been an honor and a pleasure to have my book published by the University of Chicago Press and to work with Doug Mitchell. Thanks to him for his bolstering encouragement, for selecting three anonymous reviewers who really appreciated what I was trying to do and offered suggestions for improvement, and for assembling a great team for putting the book through production. Thanks to Dawn Hall for her adept copyediting and to Mireya Herrera for designing figure 1.1, the map of behavior genetics.

Lastly I want to thank my family for all of their love and support. Thanks Mom, Dad, and Dave. But the greatest thanks, and love, go to my children Charlotte and Beatrice, and especially my wife Betsy Blanchard—my angel, muse, partner, and sometime overseer. No written acknowledgment can ever repay my debts to her, especially since they're ever accumulating. Thankfully I get to work them off in person.

Introduction

In the mid-1990s a major controversy about the innateness of human differences rocked the field of behavior genetics. In 1994 Harvard psychologist Richard Herrnstein and policy analyst Charles Murray published *The Bell Curve*, an 845-page tome about US economic inequality.[1] Herrnstein and Murray drew on research in psychology and behavior genetics to argue that US class structure can mostly be attributed to inequalities in individual intelligence as measured by IQ, that IQ is mostly an innate capacity of individuals under genetic control, and therefore differences in education and upbringing are not responsible for social inequalities. Their most provocative argument concerned race. Herrnstein and Murray claimed that genetic differences largely explain the lack of black and Latino success relative to white and Asian, though environment plays some role. The implication was that discrimination is mostly over, and that unequal social structure is genetically determined. Policies aiming to uplift minorities and the poor are doomed to fail, they claimed; instead, the "cognitive elite" must find ways to manage a permanent genetic underclass.

The Bell Curve drew heavily on the work of J. Philippe Rushton, a University of Western Ontario psychologist, for its claims about genetically driven racial differences. In 1994 Rushton pushed the racial argument much further in a book of his own, *Race, Evolution, and Behavior.*[2] For Rushton, inequality in America was but one manifestation of a universal racial hierarchy in intelligence, personality, civilizational achievement, family stability, and propensity to social order. Across indicators, Rushton claimed, "Mongoloids" came out on top, "Caucasoids" were a close second, and the hapless "Negroids" were far below. He explained this pattern in terms of evolved strategies: as ancient humans left Africa they faced harsher environments, which forced them to develop greater intelligence, sociality, and sexual restraint. One residue of this, according to Rushton, is that "Negroids" have big penises, small brains, and don't care much for their children; "Mongoloids" have small penises, big brains, and invest heavily in their children; and "Caucasoids" are somewhere in between.[3] Thus Rushton's charge was that black people are genetically and evolutionarily maladapted to modern, civilized life.

The race controversy took a twist the next year at the Behavior Genetics Association (BGA) annual meeting. Responding to the renewed attention to race and behavior, President Glayde Whitney, a mouse taste specialist from Florida State, organized a symposium called "Group Differences: Research Directions." There, Rushton and Arthur Jensen and David Rowe, both noted psychologists and race researchers, argued for the genetic reality of racial differences. John Loehlin, a psychologist, argued cautiously against race research in his talk "Group Differences—Should We Bother?" In his presidential address Whitney answered this rhetorical question with an emphatic, "yes." Whitney entreated his colleagues, on the occasion of the BGA's twenty-fifth anniversary, to undertake an ambitious research agenda to discover the genetic roots of racial behavioral differences. As justification he cited evidence that international crime rates are directly related to the proportion of blacks in the population.[4] Then he accused anyone who might deny this argument of having "marx-itis" and began to apply this label to members of the audience who had been critical of race research in behavior genetics.[5] The audience was shocked. To many the speech was a racist screed that misrepresented the field; some were embarrassed that the mostly black staff in the banquet hall had to listen to it. Several walked out in protest, including members of the BGA's executive committee sitting at Whitney's own table.[6] Yet the race question, historically the field's thorniest problem, could not be ignored.

These events drew a tremendous amount of attention to the field. Hundreds of thousands of copies of *The Bell Curve* were sold. It was the cover story of *Newsweek*, *New Republic*, and the *New York Times Magazine*. *Nightline*, *MacNeil/Lehrer Newshour*, *McLaughlin Group*, *Charlie Rose*, and *Primetime Live* covered it on TV.[7] Later, Rushton sent an abridged version of his book to thousands of social scientists and journalists.[8] Since Herrnstein had died just before *The Bell Curve* came out, Rushton became the social science authority to whom media looked for defense of the book's ideas. Both books were discussed in hundreds of articles; public forums and debates were staged. They became the occasion for a national debate about American society.[9] Even Whitney's speech to only two or three hundred behavior geneticists was covered in the news sections of *Science* and *Nature*.[10]

The situation certainly had its benefits for the field of behavior genetics. All these buzz makers had relied heavily on the field's concepts and claims to argue that racial behavioral differences are genetically determined, and thus their work raised the field's public profile. Behavior genetics seemed to have information crucial to the fate of a democratic, meritocratic society—an enviable position when most scientists toil in obscurity, struggling to explain how their research matters to people's lives. Perhaps, as some claimed, behavior genetics was simply revealing cold, hard truths about the inevitability of inequality and poverty.[11]

The spotlight was an uncomfortable one, however. Many commentators noted parallels to an earlier era's eugenics-motivated concern with the socially "unfit" and "racial degeneration." They asked: Was behavior genetics "racist science"? Was it a first step to reviving Nazi eugenics?[12] Others challenged the science, claiming it was too flawed to help guide social policy.[13] When the eminent geneticist David Botstein was asked why so few geneticists had publicly criticized *The Bell Curve*, he responded, "The answer is because it is so stupid that it is not rebuttable."[14] This was an ugly problem forced on the field. Behavior geneticists had to ask themselves: If Herrnstein and Murray, Rushton, and Whitney were misusing the science, did behavior geneticists have a responsibility to denounce them? A failure to do so might imply that behavior genetics itself was either irresponsible science or "too stupid" to warrant attention from serious scientists, as Botstein's gibe might imply.

The field's collective responses to the controversy were quite unexpected. A small cohort of behavior geneticists did publicly attack the racial arguments. Jerry Hirsch, a fruit fly geneticist, organized sessions challenging *The Bell Curve*'s scientific claims at the BGA and the American Association for the Advancement of Science (AAAS) meetings, and he later edited a slate of critical articles in the journal *Genetica*.[15] Douglas Wahlsten, a mouse researcher, wrote a scathing review of Rushton's book.[16] And in response to Whitney's speech and the BGA executive committee's unwillingness to censure him, Wim Crusio and incoming president Pierre Roubertoux, both mouse neurogeneticists, resigned their positions on the board. But all of these bold responses were taken by *animal* behavior geneticists whose research was very far from matters of human intelligence and racial differences.

By far, the field's broadest, most public response was to *embrace* the arguments of *The Bell Curve*. "Mainstream Science on Intelligence," an editorial in the *Wall Street Journal* signed by fifty-two intelligence researchers, including about two dozen leading behavior geneticists, endorsed Herrnstein and Murray's picture of IQ and rejected the common notion that they had grossly misrepresented science.[17] The statement sidestepped the genetically stratified society Herrnstein and Murray envisioned. But on the genetics of race differences, it had two points:

> 22. Most experts believe that environment is important in pushing the bell curves [for IQ scores between blacks and whites] apart, but that genetics could be involved too.
>
> 24. Because research on intelligence relies on self-classification into distinct racial categories, as does most other social-science research, its findings likewise relate to some unclear mixture of social and biological distinctions among groups (no one claims otherwise).[18]

Although stated with a degree of equivocation that would comfort any liability lawyer, this statement was perceived by outside observers as well as members of the field as an endorsement of the controversial race claims of *The Bell Curve*.[19]

Less publicly, behavior geneticists also supported Rushton. Through efforts to get him dismissed from his university and despite denunciations of his work by leading geneticists and naturalists, many behavior geneticists rallied around Rushton's academic freedom and recognized him as a legitimate researcher.[20] So too with Whitney: In the days and weeks that followed his speech, conflict erupted in the BGA over how to deal with him. Two sides emerged. Those siding with Roubertoux and Crusio felt that Whitney had illegitimately used his presidential authority to endorse a racist view, and that the BGA had to censure or expel him. The other side argued that the principle of intellectual freedom demanded that Whitney must be left alone. The intellectual freedom position won the day; nothing official was done to Whitney. But many were bruised in the debate, and a number left the BGA, including Pierre Roubertoux, its incoming president, to join instead the newly formed International Behavioural and Neural Genetics Society.[21] Whitney subsequently dove head first into white supremacist politics, contributing the forward to white supremacist David Duke's autobiography and writing for the extreme right wing magazine *American Renaissance*, before he died in 2002.[22]

❧

How should we explain these dizzying events? In particular, why have behavior geneticists embraced claims widely seen as racist science? The most obvious explanation is that apart from the few dissenters behavior geneticists believed their science justified the genetic explanation for racial differences in behavior. But this is *not* the case. As a leading psychological behavior genetics researcher explained,

> I really don't think that there are tools. . . . If I find genes for IQ, someone is going to say, go and look at it for racial groups. I think it would be completely uninformative. So, racial groups differ in frequency of a gene. They differ for the frequency for lots of genes. How are you going to say—just because within a Caucasian population, this gene is associated with a [trait]? . . . You've got no degrees of freedom when you're studying racial groups. I think, so, I don't even think the molecular genetics—I don't see how it's going to shed light on the etiology of racial differences.[23]

Despite this view, the speaker was a signatory of the *Wall Street Journal* statement. Another field leader explained a colleague's fury at the quality of Rushton's work: "I know someone, a pretty prominent behavioral geneticist, who's livid at Rushton for one of his books. . . . He thought that the analysis was just completely flawed. And this was not an ideologue. In fact, I know this person to be pretty conservative politically and [he] would probably be pretty open to the type of thing Rushton might argue."[24] These sentiments—that racial claims are not only flawed but also impossible to justify with available tools—continue to be widely held by behavior geneticists.

To be clear, one can *believe* that genes cause racial behavioral differences and also that science cannot substantiate that belief. What is more, the *Wall Street Journal* statement was really a sociological one about what some experts believe, not what the science proves. But by behavior geneticists' own definitions of scientific possibility, all this actually militates against the idea that *science* has compelled their endorsement.

Thus those who believe the genetic explanation for racial behavioral differences do so despite, not because of, their field's science. Furthermore, defending race researchers' scientific freedom opens up a contradiction: how can scientific freedom be invoked to defend practices that are beyond science?

Inconsistencies in the scientific account have led critics to charge that behavior geneticists are politically motivated. This line of argument holds that behavior geneticists have historically supported genetic claims about racial behavioral differences because they are racist or at least politically conservative. Psychological studies of science have long demonstrated the association in scientists between conservative politics and belief in racial inferiority.[25] Critics Leon Kamin and Stephen Jay Gould have argued separately that racial and class bias infect the ways behavior geneticists have interpreted their data.[26] Many historians and journalists have demonstrated the deep social and institutional ties between scientists promoting race difference claims and politically conservative activists and foundations.[27] The 1984 book *Not in Our Genes* by Richard Lewontin, Steven Rose, and Leon Kamin, all influenced by Marxism, argued that behavior genetics was part of a larger scientific interest in biological determinism that was inspired in general by the desire to justify the cultural status quo and more recently by a backlash against the revolutionary and egalitarian ethos of the 1960s.[28]

However, political motivations cannot explain behavior geneticists' actions either. First, despite the blatant right-wing views of some, there is political diversity among behavior geneticists.[29] Indeed, a strong testament to this is the frustration conservatives occasionally voice with their colleagues—Whitney charging them with "marx-itis" is but one example.[30] Second, behavior geneticists have generally tried to steer clear of politics. In response to the Whitney affair, behavior geneticist Nicholas Martin said, "The vast majority of the membership is fully aware of the polemic potential of much in our purview, and we try to avoid getting drawn into politics. To have all this blown in one evening by one insensitive person is galling, to say the least."[31]

Behavior geneticists have also disputed the charge that their work is political because it justifies fatalism about solving human problems. As one interviewee explained, "I'm sure there were very many well-intentioned people that thought that if we do research and we find that reading disability [for example] has a genetic component to it that this might imply that, you know, there should be less effort for special education or remediation. That's exactly the opposite of what we have in mind."[32] Political motives may animate some behavior geneticists (and some of their critics), but they cannot explain the collective patterns of action in this controversy, or the field's many others.

❧❧

This vignette opens up many of the key questions this book aims to address. Moving from the specific to the general: Why have behavior geneticists backed claims about the genetics of racial differences when doing so is disruptive, costly to scientific authority, and poorly motivated either scientifically or politically? Beyond race, behavior genetics has constantly been wrapped up in controversy; why is this so? How do behavior geneticists cope with controversy? How does this affect the knowledge they produce? What can controversy in behavior genetics tell us about the causes and consequences of controversy in other fields of science?

Misbehaving Science

Controversies, disputes among scientists, are an elemental component of science. Science proceeds, some would say progresses, by working through controversies, finding ways to settle them. Radically different accounts of how this process occurs still share the interest in controversies and their resolution. Traditional accounts of science understand the resolution of controversies through the application of rational methods, the testing and falsification of hypotheses, and the accumulation of empirical evidence.[33] Social constructivist accounts, in contrast, see the application of rationality as insufficient to explain the end of controversies. They emphasize instead the production of a social order: norms of conduct, signs of certified membership, standards of attestation and interpretation, and symbolic boundaries that enable scientists to trust each other's claims and to ignore ambiguities.[34] Common to these radically different accounts is the assumption that controversies *do* resolve. Indeed, the capacity to end controversies makes science distinctive, setting it apart from politics, the arts, and other spheres of culture that dwell in intractable disagreement. Scholars view this capacity to resolve controversies and deliver facts time and again as an important source of science's social authority.

Tracing the field's development from its origins in the 1950s through today's postgenomic moment, this book shows how behavior genetics fails to fit this image of science. Scientific controversy has been persistent and ungovernable. Settlements are temporary and unstable; disputes are liable to flare up again and again. Controversies tend to become entangled and spur each other along. The mid-1990s race controversies from the vignette above were but a moment in time. Race has dogged behavior genetics, off and on, for a half century—longer, if you count the field's eugenic prehistory. These disputes were prefigured in the late 1960s and '70s when some behavior geneticists argued that racial achievement gaps in education and success were genetic in origin, so social efforts to close them were doomed to fail.[35] Jumping to the mid-2000s, the racial intelligence argument got a molecular genetic makeover when neurogeneticist Bruce Lahn suggested that racial IQ differences might be due to evolved population differences in variants of the *ASPM* and microcephalin genes.[36]

Behavior genetics controversy goes well beyond race as well. Behavior geneticists' claims about intelligence and personality have long been disputed. For example, behavior geneticists David Rowe and Sandra Scarr have argued that parents don't affect their children's personality and that most schools are good enough to let children reach their genetically fixed intellectual potential.[37] Critics have charged them with genetic fatalism and ignoring how effective interventions can help children.[38] Controversy has also dogged behavior geneticists' claims about schizophrenia and other mental illnesses, drug and alcohol addiction, criminality, aggressive and risky behavior, and homosexuality.[39] For example, claims that criminal behavior is influenced by genes have led some to worry that genetic tests could be developed to pretreat the criminally inclined, while defense lawyers have attempted to claim genetic predisposition as mitigating their clients' culpability.[40] Similar debates followed geneticist Dean Hamer's claim to have found a genetic marker associated with male homosexuality.[41] Perhaps this would lead people to accept homosexuality as "natural" and not an "immoral choice," but perhaps it would lead to genetic tests and the selective abortion of "gay fetuses."[42]

Beyond particular behaviors, the field has engendered disputes about broad theoretical questions: Can human behavior or character be reduced to the action of genes? How much do genes constrain human capacities to change? Is it best to understand behavior by studying animals under laboratory control or humans in all their complexity? Which scientific disciplines are best suited to understand behavior (and therefore most deserving of funding and attention)? Intense, often vitriolic, debate has accompanied these questions.[43]

Adding to the perpetual conflict, behavior genetics controversies have been entangled in broader political and social disputes. Behavior geneticists have found their ideas at the heart of debates about the nature of responsibility. If genes substantially affect aggression, addiction, or intelligence, are murderers, addicts, or failing students responsible for their actions? Are social institutions responsible for seeking to improve people's lives, or are they a waste of money? Behavior genetics has also been drawn into arguments about inequality and social order. What is a fair distribution of opportunities and rewards in society? One view is that if human abilities are largely genetically determined, then maybe the rich are rich and the poor are poor because of meritocratic sorting. The other is that if human abilities are substantially caused by environment, then social hierarchies reflect capricious and unfair social processes. In contrast to sociologists and cultural critics who approach behavior genetics controversies in terms of their social implications and cultural reception, I see them as illuminating the crucial relationship between the social structure of scientific communities and the knowledge they are able to produce.

Behavior genetics is a prime example of what I call *misbehaving science*. In misbehaving science, controversy is persistent and ungovernable. Controversies wax and wane, sometimes they emerge explosively, but they never really resolve and always threaten to reappear. Scientists often work valiantly to manage them but find the contentious scientific issues irresolvable. In misbehaving science, controversy is passionate and political. Efforts to calm tensions instead often inflame passions. Scientists are confounded in efforts to draw boundaries between politics and science. If science is like a machine for resolving controversies, in misbehaving science that machine is broken.

Misbehaving science is due to relative social disorder within science. It is a situation where boundaries between science and nonscience cannot be drawn successfully. In misbehaving science norms of scientific conduct and knowledge cannot be established or enforced, or they are not accepted and internalized. However, it is not a condition of total chaos. Rather, misbehaving science is a product of partial anomie. Where scientific norms and standards are ambiguous, underdeveloped, or inappropriate to the situation, misbehaving science reigns.

Historical and sociological studies of science have recognized that *temporary* disorder with scientific fields is common. Indeed, the rich tradition of controversy studies can be thought of as analyses of breakdowns of social order where presumed boundaries between science and politics or culture are perforated, explicit and implicit standards of evaluation are uncertain, and the scientists' credibility is in doubt. Many trace how order is produced or restored so that true scientific claims could be successfully asserted—and "nature" or "rationality" do not foreordain the outcome.[44] Others examine the identities and practices that have evolved over time for the ongoing production of scientific stability and the management of disorder.[45]

Yet in misbehaving science the disorder is ongoing. Scientists cannot fully restore order, and, as is the case in behavior genetics, controversies accumulate and interact, disordering the field in different ways over time. Coping with the disorder of controversy becomes a perpetual moving target. Misbehaving science is also an ongoing crisis of authority. Following Pierre Bourdieu, whose perspective on social action guides this book, I see disagreement as endemic, even essential, to scientific fields. Indeed, disagreement is a precondition of scientific authority; successfully working through disputes accrues authority to the winners as well as the field overall. Norms are less a product of consensus than the accumulation of implicit rules "imposed" by the authoritative winners of previous scientific struggles. Part of this authority is the imposition of some forms of "censorship"—rules about holding in abeyance some forms of disagreement, for example, about certain basic scientific assumptions or about the combined social and scientific nature of competition. Thus misbehaving science occurs when the winners lack the authority to assert hegemony over the rules of the field.

Misbehaving science is not the same as scientific misconduct, that is, data falsification, plagiarism, selective reporting of results, or biased interpretations of data. Scientific misconduct involves violation of the usually informal, but nearly universally acknowledged, norms of scientific practice. Scientific misconduct is, at heart, about individuals breaking rules. Misbehaving science, in contrast, is about the ambiguity of rules, the collective lack of appropriate rules, or a shortfall in their assertion and policing. Misbehaving science is linked to confusion or irreconcilable conflict about definitions of good science, scientific recognition, field membership, or scientific responsibility.[46]

Misbehaving science and scientific misconduct are not unrelated, however. For one, a situation of relative normlessness or ambiguity around one set of issues can cloud standards of scientific practice and judgment, leading to misconduct. And, of course, as sociologists of science have long argued, the norms of proper scientific conduct are, in practice, much more ambiguous and difficult to define than scientists typically want to admit.[47] But equally important, a situation of scientific misbehavior, where norms are clouded, can lead scientists to accuse or suspect each other of scientific misconduct because that language is the root and vernacular of scientists' moral vocabulary and outlook.

By the same token, I speak of misbehaving *science* not misbehaving *scientists.* My analysis does not focus on individual "deviants." Indeed, the notion of deviants, like misconduct, implies a consensual normative order, the opposite of misbehaving science. Further, my analysis is not particularly interested in questioning or explaining the psychological, moral, or political motivations of individual scientists—for example, what, in this day and age, would possess a scientist to doggedly pursue evidence of blacks' purported genetic inferiority? Rather, I'm interested in how such a pursuit is possible at all, becoming accepted and, to a degree, rewarded as scientific conduct within a field.

⁂

This book illustrates misbehaving science through a historical analysis of how controversy has shaped the development of behavior genetics. It explains: why behavior genetics is misbehaving science, the reasons controversy has been so persistent and ungovernable, and how the political and scientific dimensions of these controversies became impossible to disentangle. Further, it makes clear why some behavior geneticists and their allies have so often willfully courted controversy and how others have coped with the situation. And, crucially, the book shows how controversy and coping with controversy have affected the production and communication of knowledge about the relationship between genes and behavior.

[handwritten margin note: What the book discusses]

Many have written about behavior genetics, though usually either to praise or criticize its science or to debate its political, social, or cultural impacts.[48] I approach it from the sociology of science—my aim is to explain how the context in which behavior geneticists work affects the knowledge they produce. *Misbehaving Science* emphasizes the practical exigencies behavior geneticists face in coping with controversies. The controversies on which I focus demonstrate how scientific disputes with a strong political valence can produce chaotic conditions through the particular pressure they exert on the boundaries, authority, practices, relationships, and identities of behavior genetics. I show that behavior geneticists' actions during these controversies are largely pragmatic responses unified by the desire to secure room for maneuver for those working under the label of "behavior genetics." Their actions are differentiated, however, by the fact that they occupy different positions that give them conflicting visions of behavior genetics and its possibilities.[49] Scientific and political motivations matter, but rather than acting independently they are mediated by the practical demands of coping with controversy and the evolving state of the field.

Misbehaving Science shows how controversies, and scientists' responses to them, have successively remade the field of behavior genetics. This process has had two decisive patterns. The first has been the long fragmentation of behavior genetics into an *archipelago* of subfields that are loosely integrated, normatively differentiated, and uneasy about mutual identification and oversight. The second has been the field's perpetually problematic scientific authority (the inability of behavior geneticists to fully convince other scientists of their scientific bona fides) and ambivalent position in scientific status hierarchies. To summarize the story crudely: The first condition produces the situation of relative anomie in behavior genetics, and the second produces an interest in scientific provocation and risk taking among many behavior geneticists. Social fragmentation and anomie have had intellectual consequences: Research has been locked into furrows, and it has often proceeded by repeating established procedures; widely denigrated concepts and interpretations (about race and genetic determinism, for example) have enjoyed long lives; and what counts as behavior genetics has been narrowly defined, alternatives have been seen as antagonistic, and potential syntheses have been blocked. We will see that available knowledge of the relation between behavior and genes might have been very different had behavior genetics not fallen into misbehavior.

This project draws from a variety of empirical sources to reconstruct behavior genetics' development and structure. Most important are thirty-six in-depth interviews I conducted with prominent members and critics of behavior genetics. I selected interviewees with two criteria in mind: first, for their knowledge of and experience with the field's long history and many controversies; and second, to maximize the coverage of positions in the field in terms of the different research traditions and disciplines in which they work.[50] I also draw from fieldwork that consisted of attending professional conferences, a week-long summer training course, and a series of meetings devoted to bringing together behavior geneticists and their critics to promote "public understanding" of the field. In addition, I use the published record—from scientific papers and reviews to commentaries, letters, and historical reflections—to reveal the progression of substantive concerns as well as key information on members' aspirations, relationships over time, and the transpiring of controversies. I have also analyzed a variety of data, like scientists' CVs, published surveys of behavior geneticists, grant databases, and publication acknowledgments to gain information on the scientific interests, training, and funding of behavior geneticists.

By using a variety of data sources and especially sources from individuals that occupy a range of positions in the field, I have been able to triangulate their perspectives and thus reconstruct the stakes of struggles that define behavior genetics. All the conflict characterizing the field's development, and participants' willingness to talk about it, has made easier the analytic job of finding points of disagreement and identifying their significance, distinguishing the variety of motives inducing participants' actions, and denaturalizing the prevailing state of affairs by identifying alternatives to what exists presently. But it has presented difficulties as well: on some issues, the concept of heritability, for example, there is so little common ground that it becomes difficult to articulate the stakes of disagreement. I have attempted overall to adopt a "naturalistic" stance on key contested issues of behavior genetics, such as the field's definition, its scientific essence, the meanings of its concepts, its norms of conduct, and so forth.[51] Thus I do not invoke criteria from outside the frame of action to adjudicate who is "right" and "wrong." Instead I triangulate the perspectives and articulate their differing claims and (often tacit) warrants to reveal the character and bases of the conflicts that characterize behavior genetics. The context- and conflict-specific range of meanings of these empirical objects presents a writing, thinking, and reading challenge because it is usually important to keep both a specific range of meanings and the generally contested character of any of these concepts in mind even if the shorthand label is all that appears on the page.

Plan of the Book

The first chapter of *Misbehaving Science* lays the groundwork for analyzing behavior genetics and establishing its broader significance. Misbehaving science is persistent ungovernable controversy linked to the disorganization of normative order in a scientific field. Drawing from the
sociology of scientific knowledge and Pierre Bourdieu's theory of social
fields, I lay out an analytic strategy for studying scientific anomie by em-
bedding it in an account of the evolving organization of scientists' competition for recognition and resources. Then I develop the concept of misbehaving science comparatively. I demonstrate that controversy might be
organized differently in behavior genetics by showing how related fields
have managed the problem. Nor is misbehaving science unique to behavior genetics; comparison to several other fields suggests it is a growing
problem in science. Finally, through a comparison to the field of neuroscience, I offer a baseline description of the social organization of behavior genetics since the mid-1990s race controversy. Behavior genetics is
disciplinarily fragmented—I call it an "archipelagic field"—and its status
and resources are problematic. These are the basic conditions that lead
to its affliction with misbehavior.

Each subsequent chapter tracks a controversy that has been a crucial point in the development of behavior genetics. Rather than focus on
disputes about particular behaviors, each chapter considers a moment
when a crucial intellectual or technical debate was linked to a practical or structural change in the social or cultural organization of the field.
These are also public controversies: in different ways, at stake in each
controversy are the social implications of behavior genetics or its boundaries and relationships to other scientific and nonscientific fields.

Chapter 2 starts in the 1950s when scientists worked to establish a
new field devoted to the genetics of behavior. As they began to assemble
a network and a research agenda, they were deeply concerned with the
legacy of eugenics and racism that had long tainted the topic. Their aim
was to create a unified, disciplinary field that would avoid the controversial, political potential of behavior genetics. Thus they explicitly tried to
avoid those topics and the older generation of researchers who had advocated them. Further, they tried to create an intellectually heterogeneous
field scientifically secured through vigorous interdisciplinary debate. Researchers using both animals and humans formed an implicit pact: they

would defend and police each other, and the synthesis of their perspectives would keep behavior genetics scientifically dynamic and rigorous while preserving its responsible social relevance. But behavior geneticists' efforts to secure their young field from controversy would not last.

Chapter 3 traces how behavior genetics was consumed by controversy when in 1969 psychologist Arthur Jensen claimed that intellectual differences between blacks and whites are genetically determined. I show how behavior geneticists came to embrace Jensen and his claims—a direct inversion of the field's position from just a few years before. I argue that the reasons were neither scientific nor political, but pragmatic. Doing so seemed the only way to weather the firestorm of controversy and defend behavior genetics from its many critics. I show how these events began the fragmentation and intellectual narrowing of the field as many geneticists, biologists, and social scientists began to dissociate themselves.

Chapter 4 considers an abiding controversy that sharpened after the race controversy—is it best to study the genetics of human behavior in humans or animals? Animals seemed to have advantages in the ways they can be experimentally manipulated, but how to link their behaviors to humans' is often difficult. I show that this intellectual debate was tied up in conflicting social visions of behavior genetics as a field. In the aftermath of the race controversy, the field had become fractured, distrustful, and closed to criticism. A bunker mentality prevailed. Animal researchers thought their work could help reestablish the field's scientific bona fides, but to link animal to human research would require investing in a social and cultural organization that promoted trust, integration, and openness. Human researchers, in contrast, saw behavior genetics as a scientific identity they could use to compete in their home disciplines. Thus they were relatively uninterested in the robust vision of behavior genetics as a field. As the human researchers gained power, the vision of behavior genetics as a quasi-discipline foundered, and its anomic, archipelagic form solidified. In the process, the intellectual definition of the field narrowed: there was less theoretical reflection on ways to conceive the genes/behavior link, and heritability estimation supplanted the other potential ways of defining behavior genetics.

The central controversy in chapter 5 is about reductionism: to explain behavior, must scientists seek to reduce it to biological action (brains, neurons, genes), or is behavior an emergent characteristic inexplicable in terms of its elements? Before the IQ debate behavior geneticists had sought to empower the reductionist perspective by integrating diverse scientific perspectives. But the IQ controversy had called behavior geneticists' scientific tools into question, brought them into conflict with other scientific views, and weakened the institutional basis of their field. The "social order" perspective in the sociology of science would anticipate that behavior geneticists would try to valorize their science by strengthening the field and settling disputes about it. I show, in contrast, that during the 1980s they advanced reductionist ideas by capitalizing on the field's flexibility and normlessness and by provoking and sustaining controversy with their opponents. Disagreement persisted about their reductionist claims, but it also spurred practical scientific activity that sustained behavior geneticists. Intellectually these disputes drove behavior geneticists to conceive behavior in a static, asocial way and also to interpret their findings deterministically.

In the 1990s people thought the advent of molecular technologies spurred by the Human Genome Project would revolutionize behavior genetics. Chapter 6 considers the controversy that has unfolded since then when two possibilities for behavior genetics were debated: Some believed technology would finally resolve controversies and integrate the field by allowing direct links between behavior and DNA and by drawing expert geneticists. Others worried that it would encourage more controversial genetic determinist claims. I show that neither happened. Molecular methods failed to deliver the expected results; as a result, the older behavior genetic techniques became a relative oasis of certainty. Further, as molecular geneticists entered behavior genetics they became yet another island in the archipelago. Faced with this new competition, the behavior geneticists began pursuing research that positioned them as holist and environmentalist—thus reversing the way they had positioned themselves earlier.

Chapter 7 also covers the 1990s then moves to the present in considering the dispute about behavior geneticists' responsibility for the public communication of their work. Behavior geneticists and their critics have both been unhappy about the often-sensationalistic way the science is publicly portrayed. Behavior geneticists have tended to blame the media and public ignorance, while critics have implored scientists to greater public responsibility. I focus on the mediating influence of the field to show how "responsibility" and "irresponsibility" become socially organized. First, I show how controversies have led field members to define responsibility in terms of intellectual freedom and participation in public communication forums. Next, I show how the archipelagic structure of the field has made it impossible for practices of mutual commitment and mutual regulation that might help inculcate responsible communication to emerge. Further, among the key symbolic resources available to participants is the opportunity to be provocative and acquire scientific notoriety; irresponsible practices have been rewarded even if they're not sanctioned. The chapter concludes by showing some of the social and cultural effects of behavior genetics' brand of charged, geneticized discourse.

The book concludes by reflecting on the broader significance of misbehaving science. After describing some of the study's analytic contributions for thinking about scientific fields and controversies, I suggest that structural tendencies similar to those promoting misbehaving science in behavior genetics are common in contemporary scientific fields. I end by showing how the idea of misbehaving science helps us to think critically about popular ideas for the management of contemporary science and the knowledge produced by behavior genetics.

CHAPTER ONE

Studying Misbehaving Science

Describing behavior genetics as *misbehaving science* is to claim that its propensity to experience persistent, unresolvable, often politicized controversy is linked to its history, organization, and practices. Controversy, in other words, could be otherwise: temporary or arranged differently. The persistence of controversy in behavior genetics is due to social fragmentation and anomie as well as its liminal position and ambivalent status at the interstices of other disciplines. Thus the story of behavior genetics, and misbehaving science more generally, is wrapped up in social structure, more specifically in its constitution as a semiautonomous social space with its own boundaries, rules of action, and patterned struggles for resources. To understand any misbehaving science we must approach it as a field in the sense social theorist Pierre Bourdieu proposed.[1]

In his theory for analyzing the dynamics of social fields, Bourdieu argued that all fields have a dual structure; each is simultaneously a field of forces and a field of struggles. In science, the field of forces can be thought of as the set of implicit "rules," both intellectual and social, governing how scientists can make legitimate claims and garner success. This is to say, how they can gain "scientific capital" defined "inseparably as technical capacity and social power . . . to speak and act legitimately . . . in scientific matters."[2] These capacities are ultimately social; they are dependent on the scientist receiving the implicit or explicit recognition of her peers who, collectively, are the arbiters of what counts as legitimate knowledge in a particular corner of the scientific world.

The field of struggles then, is the competition that each scientist must engage with her peers to force them to recognize her scientific contributions as legitimate. As her cocompetitors they are disinclined to offer this recognition lightly. These struggles are twofold: Scientists simultaneously endeavor to convince others that they are *doing* "good science" that is worthy of recognition, but they are also seeking to *define* good science in a way that will benefit them by valorizing the kind of science they do. A scientist might seek to do more research in the image of what has come before and thus reinforce prevailing definitions of science, or she might try a new synthesis or analytic strategy with the aim of overthrowing the prevailing definitions and establishing a new one. The outcomes of these ongoing struggles redistribute scientific capital and affect the rules for accumulating it, which is to say they transform or conserve the field of forces that govern subsequent struggles.

To reconstruct the field is to reveal its "logic of practice"—the way the forces and struggles organize actors' perceptions, interests, and strategies depending on their investments and positions in the field.[3] The language of "strategy," "logic," and "investment" can be misleading. In the Bourdieusian framework it refers less to explicit, self-conscious calculation and more to the implicit, practical reason that flows from an embodied sense of how to play the "game" according to evolving tacit rules organizing a space. Bourdieu's concept of the illusio—the sense that playing the game is "worth the candle"—is meant to remind us that the analysis in term of actors' "strategies" for seeking "recognition" is a post hoc reconstruction of what, for actors, is the experience, in this case, of doing good science.[4] Further, the language of "rationality" and "interest" should not be understood as appealing to some universal calculation of benefit for actors. What counts as valid "interests" to pursue or "rational" (or reasonable) means to pursue them are given by the rules of the field.

Analyzing misbehaving fields in this way enables us, first, to steer out of the deep ruts of internal and external explanation that often typify analyses and critiques of controversial sciences. Analyses of behavior genetics have tended to explain the field's ideas either in terms of their internal logic and evidence or the external political motives animating them.[5] This divide is mirrored in the debate between the field's critics and its advocates: each side tends to describe their own work scientifically and their opponents' work as politically tainted. In contrast, the field approach analyzes the shaping of scientific claims and the challenges they've faced in terms of the social relationships—the forces and struggles—that make up the field. This approach doesn't deny internal or external forces, but insists that they're mediated by the relationships and practices comprising the field. We have to account for how one rather than another set of scientific arguments gains salience within the field and how various political motivations are legitimated or delegitimated as "scientific" action. The field approach asserts that we cannot understand why scientists do what they do, and explain the knowledge they produce, without understanding their competition for mutual recognition at the mesoscale between the microlevel of scientific practices and the macrolevel of extrascientific contexts.

Second, this approach enables a dynamic and historical definition of controversial fields. Fields are the result of a set of struggles over boundaries and definitions rather than the unfolding of a unified idea. Thus "behavior genetics" is not the ancient interest in the genetic inheritance of behavior, nor is it a coherently defined type of scientist.[6] Rather, it is a heterogeneous and dynamic set of individuals and subgroups in competition with each other. They share some motives and aims but certainly not all. Who counts as a behavior geneticist changes as boundaries, affiliations, and identities shift. And as I will show in subsequent chapters, sometimes scientists are struggling *not* to have themselves or their ideas counted as "behavior genetics."

With this understanding, it isn't necessary to come up with some factor common to behavior geneticists to explain seeming anomalies. If race differences claims persist, for example, this doesn't mean that behavior geneticists are secret racists (or in false denial). Rather, the balance of forces in the field makes it beneficial for some to make these claims and impossible for others to urge their elimination. This perspective allows us to steer out of another related set of ruts: the reification of opposing camps in a science. In the case of behavior genetics this enables me to avoid the assumption that conflict over the field is the product of an eternal battle between "hereditarians" and "environmentalists."[7] We can see, instead, that these categories are themselves historically changing positions or identities that scientists redefine over the course of their mutual competition.[8]

Third, the field approach reinforces seeing controversy in terms of scientists' efforts to seek recognition. Scientists in a field, though all oriented toward science, have different motivations and strategies for pursuing it. The differentiating effects of this competition suggest that scientists can have different interests in publicizing transgressions as tools to seek recognition or deny it to others.[9] A scientist might advance a race differences claim to gain attention by shocking audiences and projecting bravery at violating a taboo. Or a scientist might charge another's ideas as "scientific racism" to discredit them politically and scientifically. It is common to see "politics," "the media," "publicity," or the like as obvious distortions of science. Following the field approach, we should see them instead as potential resources in scientists' struggles for recognition. For example, how do scientists exchange media publicity for scientific capital in behavior genetics? How do they use charges of "politicization" to redefine what constitutes its scientific capital? In other words, how does the logic of the field define legitimate and illegitimate "political" (or "economic," or "media") interests?

The field approach also enables us to see criticism of behavior genetics not only as scientific arguments but also as *relationships*. Behavior genetics is embedded at the interstices of several scientific fields—genetics, psychology, neuroscience, psychiatry, and biology among others—competing to gain, and sometimes to cede, jurisdiction.[10] Affiliations with behavior genetics have varying connotations in these different neighboring fields and thus affect the kind of recognition scientists can hope to receive.[11] Arguments about behavior genetics therefore concern scientists from other fields seeking to extend or block behavior genetics' authority in accordance with their own struggles for recognition.

In addition, scientists' different scientific aims can also give them interests in different versions of relative "social order." For example, we will see how a major conflict among behavior geneticists has concerned whether the field should be a tightly organized field with inwardly directed mutual recognition or a technical subfield whose researchers primarily seek the recognition of those in other fields. Summing these efforts together, the field approach allows us to see controversies not simply as truth contests, as they are traditionally viewed in the sociology of science, but as accumulating, interpenetrating "stages" in the reorganization of behavior genetics' fields of "forces" and "struggles."

Here it is prudent to address potential tensions between field theory and two concepts central to my analysis. First, the idea of a field governed by a common set of forces and struggles seems to be in tension with the metaphor of an archipelago of loosely connected islands. I approach field theory as an analytic strategy for understanding the shifting boundaries and structures governing a social space and not as a substantive image of a particular kind of social space. Field analysis helps us see how behavior genetics became archipelagically fragmented and how this affects the different ways forces and struggles organize the various islands. Indeed, it is precisely behavior genetics' *ambivalent* "fieldness" that explains many of its distinctive practices and problems, and the paradox of consensus that behavior genetics is important but utter disagreement about why or even what behavior genetics is.[12]

Second, the idea of anomie—central to the idea of misbehaving science—might seem to be at odds with the field framework. In Émile Durkheim's social theory, anomie is a situation where a community lacks sufficient or appropriate norms to govern and integrate its members' behavior.[13] Bourdieu's field theory, with its focus on conflict and competition, eschews the notions of community and commonality that underwrite Durkheim's view. Therefore, anomie might seem an inappropriate concept in this framework. However, Bourdieu's competition isn't anarchic, Hobbesian struggle, nor is it crypto-economism or rational actor theory, as is sometimes charged.[14] Rather, Bourdieu shares a view explained by the philosopher Charles Taylor that individual agency and the capacity to compete depend on substantial, shared preunderstandings.[15] Fields are not just arenas, but are fields of forces: explicit, implicit, and evolving rules and unthought assumptions, or *doxa*, governing competition.[16] Viewed this way, a field can certainly face anomie. It is not general normlessness and lack of community integration, but the institutionalization of "rules that are the lack of rules." As we will see, some of behavior genetics' "rules against rules" act subjectively, as when behavior geneticists discourage each other from "airing dirty laundry." But others are objective, as when disconnections between intellectual traditions, narrowing definitions of legitimate science, and engagement with scientific audiences outside the field in preference to engaging other behavior geneticists, all de-intensify the competition within the field. I thus conceive of anomie in terms of Bourdieusian rules of the game—objective and subjective structures produced by struggle, principles of vision and division—not shared norms and community integration.

Fields and Structured Controversies

My field analytic approach to controversy in behavior genetics straddles two important traditions in the social studies of science. First, studies of controversy and knowledge production have long been a mainstay. They have been analyzed as episodes of substantive dispute among scientists. Social and historical analysts have found them valuable as moments that lay bare the processes driving knowledge in formation. The focus of this tradition has been to explain knowledge production mainly in local, practical, and substantive terms.[17] How were social relationships, meanings and interpretations, and technical practices organized so as to make this, rather than that, fact hold true?

In behavior genetics many of the substantive disagreements—about the genetic effects on intelligence, personality, mental illness, crime, and so on—have a similar pattern of polarization.[18] Rather than focus on them, I consider controversies where the substantive dispute at stake became entangled in a reorganization of the struggles and forces that constitute the field. In other words, I use a particular set of controversies to tell, following theorist Anthony Giddens, the "structurational" history of the field.[19]

Approaching behavior genetics controversies as moments of restructuring allows the analysis of knowledge production at higher scales than the local contexts favored in much science studies.[20] It requires, first, that we map the relevant intellectual commitments to the scientists competing to define the field and in terms of the implicit, evolving "rules" for succeeding as a scientist. What are the methods, forms of argumentation, facts, and interpretations of results that prevail as authoritative among particular scientists over time? These commitments shade into a second level, the styles of knowledge production that evolve over time. Behavior geneticists debate, for example, whether it is more important to develop analytic tools that are specially tailored to the particular problems of linking behaviors to gene action or tools that are portable and that can be made relevant to various behavioral sciences; whether scientists should tackle multiple dimensions of one behavioral problem or apply their tools to many different behaviors; whether behavior genetics should be advanced by attacking competing paradigms or by forming diverse intellectual partnerships. And at a third level, how is the space of intellectual possibility structured? How does the structure of scientific competition cultivate some kinds of intellectual work while rendering others nearly unthinkable? In behavior genetics we will see the effects of two crucial patterns: first, gradual fragmentation has severed connections between different kinds of scientists, leaving potential lines of research unpursued; second, due to weakness in certain forces encouraging competition and criticism among behavior geneticists, researchers observe a laissez faire ethos yet still follow fairly narrow paths.

The other main sociology of science tradition this project intersects is the institutional analysis of specialty formation and growth. These studies ask how groups of scientists are able to constitute themselves into stable, productive communities, and the studies tend to focus on scientists' efforts to rally each other, secure patronage, institutionalize themselves (in the university or other organizations), reproduce the community through training, and cope with problems of generational succession.[21] My analysis of behavior genetics draws on these ideas to demonstrate a crucial condition under which behavior geneticists work—namely, the field's weakness: its somewhat tenuous grip on material resources, institutionalization, and legitimacy, and its ambivalent status within academic status hierarchies.[22]

But my field development through controversy framework differs from the specialty formation tradition in two main ways. First, the specialty formation tradition tends to assume that specialties are communities that are unified by common scientific purpose and values whatever their disagreements. Behavior genetics is not a community in this sense; it is a conflict among scientists over definitions and boundaries and, as we will see, this makes "behavior genetics" highly unstable over time. At an empirical level, this makes "objective" measurements of favored specialty formation variables (funding, membership, number of publications, citation networks, and such) impossible. Any such efforts will automatically favor the definitions of certain parties at particular times. When I present such information I try to be specific about *which* behavior genetics they represent.

Second, the specialty formation literature threatens to miss what is distinctive about controversy in behavior genetics. For most specialties the formation process is one of overcoming controversies: Most basically, what is this specialty's raison d'être? But also, what is its jurisdiction? And how can it establish legitimacy and secure resources to stabilize and grow? At the field's beginning, as we will see, behavior geneticists followed this playbook. But soon they discovered that *promoting controversy*—provoking scientists and the public, cultivating publicity and notoriety—were viable, if dangerous, tools for procuring resources, attracting attention, and building careers. *Misbehaving Sci-*

ence is, in part, about how controversy, more than a problem to be overcome, can be a tool of field formation and scientific growth.

From Naturalism to Critique

My analysis of behavior genetics spans an apparent contradiction. On the one hand, I approach the science naturalistically and symmetrically, but I also consider it critically. How is it possible to suspend judgment and describe, but then to judge? First some background: Since David Bloor's *Knowledge and Social Imagery*, most sociologists of scientific knowledge have approached their topic symmetrically and naturalistically.[23] That is, they have attempted to explain epistemic success and error in the same terms, and they have sought to employ the criteria of belief mobilized by actors themselves, rather than importing standards from outside the frame of action. I, too, follow Bloor's approach in analyzing behavior genetics. My role isn't to criticize or champion the scientific arguments or findings of behavior genetics or to argue about the negative or positive social implications of its research.[24] Nor is it to show genealogically that its current ideas are tainted by the problematic views of the past—namely, eugenics and scientific racism, or to argue that it has transcended them.[25] That is, I don't pick sides in the controversies about behavior genetics, but rather seek to explain the dynamics that give rise to them and set the terms of their competition.

Craig Calhoun has characterized critical theory as "the project of social theory that undertakes simultaneously critique of received categories, critique of theoretical practice, and critical substantive analysis of social life in terms of the possible, not just the actual."[26] Naturalism and symmetry open up a specific range of critical possibility: unmasking and denaturalizing a process of construction that is usually misrecognized as given by rationality and nature.[27] My field approach to behavior genetics takes this form of critique as a starting point but pushes further, mapping onto Calhoun's definition. As I described earlier, field theory breaks with received categories and theoretical practice, for example, by transcending explanations of behavior genetics in terms of hereditarianism versus environmentalism and internal versus external analysis. That is, my aim is partly to critically revise the way we understand behavior genetics and the well-worn approaches toward it by rethinking scientific controversies, how they affect the development of scientific fields and the production of knowledge. Pushing further, by uncovering the historical and structural limits and possibilities of the field in terms of the knowledge production it supports, I aim to show how scientists' knowledge, or claims to knowledge, about the relationships between genes and behavior could have been different had the field developed in other ways. That is, I show that what is possible to know, or might have been possible, is much wider than what is actually known.

A field analytic approach does not judge particular claims of social actors as good or bad, true or false, but it does admit of a sociological criterion of defining better and worse spaces of scientific practice. Successful science broadens intellectual possibilities and enables scientists' critical reflexivity; failing science prevents them. This criterion of scientific success or failure—even of better and worse knowledge—does not violate the principles of symmetry and naturalism because it is not about importing epistemological or political criteria to pick sides in scientific controversies. Rather, it targets a higher epistemic scale by considering the properties of the social milieu in which a controversy takes place. The field approach involves reconstructing a logic of scientists' practice, which goes beyond naturalism; it is an explanation of the game of social action, its capacities and constraints, that is initially not available to the players of the game, at least not in the terms in which they play it. However, this understanding can become available to the field and, in principle at least, become reflexively incorporated into practice.[28] Thus field theory is a critical approach that aims to go beyond denaturalizing and unmasking scientific practice, and it aims, ultimately, to improve science. —▸ Why use field theory.

Behaving and Misbehaving Science

Misbehaving science is affliction with persistent ungovernable controversy. To what extent is this condition distinctive to behavior genetics? Compared with the fields it overlaps substantively, behavior genetics is uniquely afflicted. Yet misbehaving science is a condition that may be widespread but little recognized in science.

A widely held view is that the combination of genetic claims and socially important behaviors makes behavior genetics inherently controversial. Many of the scientists I interviewed expressed this view.[29] Philosopher Erik Parens agrees that controversy infects conversations about behavior genetics because basic questions of human identity, social responsibility, and free will combine with disciplinary differences about the meanings of concepts.[30] And philosopher James Tabery has argued that clashes among scientists are the product of metatheoretical disagreements between the biometrical and developmental research traditions.[31] If passionate controversy is inherent to behavior genetics' cultural interface or the intellectual paradigms at play, then perhaps it isn't socially mediated misbehavior at all.

But other fields that occupy the same fraught territory at the intersection of biology and contentious political issues have been much more successful at managing and minimizing the potential for explosive public controversies. For example, historian Daniel Kevles has shown how geneticists, especially in the years following World War II, collectively worked to dissociate their work from political advocacy of eugenics— which had been deeply tainted by its association with Nazism.[32] Kevles argues that many of eugenics' conceptual and political underpinnings continued to animate geneticists, but they no longer publicly sold their field on the promise of improving the human gene pool. Instead of coercive public control of reproduction, geneticists collectively endorsed principles of individual choice and nondirective genetic counseling. Rather than dubious behavioral traits like feeblemindedness and pauperism they focused attention on clear genetic diseases. Edmund Ramsden has told a similar story about the field of demography: at about the same time demographers were pushing eugenics out of their field by focusing their research on population quantity, they were turning away from assertions about population quality, which were at the heart of eugenics.[33] William Provine has traced geneticists' progressive rejection during the twentieth century of claims that racial groups have hereditary mental differences and that miscegenation is harmful.[34] He argued that this move had much less to do with the accumulation of data about race than the pragmatic desire of geneticists to dissociate the field from this social and political controversy.[35]

Geneticists' collective efforts to manage controversy are not limited to matters of eugenics, race, and behavior. The early 1970s experiments by biochemist Paul Berg and others to manipulate and recombine viral DNA from one organism to another prompted fears about the safety of the technology. Several municipalities where much DNA research was conducted threatened bans. Geneticists responded to the controversy by holding several important meetings to discuss the risks and craft guidelines for safe recombinant DNA research.[36] Geneticists have long sought to manage controversies through the collective reordering of the field.

The field of neuroscience also demonstrates that linking biology and behavior need not produce the kind of controversy behavior genetics has faced. Like behavior geneticists, neuroscientists study cognition and intelligence, sexuality, mental disorders, alcohol and drug use, and propensities to aggressive and risky behavior, all traits imbued with intense cultural anxieties. Yet neuroscience's most fraught controversies have involved, on the one hand, protests of animal experimentation by animal rights activists, and on the other, fairly academic disputes about reductionism; for example, is brain chemistry sufficient to explain human motives or actions?[37] Neuroscientists have basically avoided the kinds of contentious, politically weighted pronouncements—particularly racial comparisons—that have made behavior genetics infamous.

Clearly, the territory that behavior geneticists plow is fertile with the potential for controversy. The topics they study are matters of intense cultural concern, and the tools they use seem to have irreconcilable conceptual properties with those other intellectuals favor. But in itself, this situation need not consign behavior genetics to persistent, ungovernable controversy. Other parts of genetics and other fields linking biology and behavior have not experienced the same kinds of problems. Members of these fields have responded to actual or potential controversies differently. Confronted with problems, their collective responses have successfully managed them, which is to say controversies have been ended or not allowed to explode in the first place. These fields seem to have arrangements that discourage individuals and groups from being publicly provocative. Indeed, they raise certain questions about behavior genetics: Why has it focused on behaviors that are destined to provoke controversy? Perhaps Tabery is right that the biometrical and developmental research traditions have conceptually irreconcilable ways of framing the genes/behavior relationship. But why is the former linked to the field of behavior genetics and the latter to its enemies? Arguably, the paradigmatic differences are just as large between social neuroscience and neurophysiology, for example, but they are part of the same field. Why isn't behavior genetics similarly an intellectual big tent that houses an array of competing paradigms? Behavior genetics' pattern of and propensity to controversy could have been different; why did the field develop in this way?

꿏ᢤᢣ

While behavior genetics is an outlier as misbehaving science, misbehaving science afflicts other prominent fields, and there is reason to think it is on the rise. The climate change field has the most obvious record of persistent, ungovernable controversy. Historian Naomi Oreskes has argued that climate scientists share a degree of consensus about the fact of average global warming and its human causes that is nearly unheard of in science.[38] Her book with Eric Conway on climate change's scientific deniers has uncovered a coordinated, well-funded conspiracy to sow doubt about global warming (among other fields) in the service of pro-business, antiregulatory interests and ideologies.[39] These analyses show the inability of the consensus to rein in these rogue doubters and the dramatic negative impact on the field's authority. The public controversy about climate change is sustained, in part, by the ungovernable scientific controversy.

The problem climate science faces is not simply well-organized political opposition from the outside. Evolutionary science faces similar opposition from religious fundamentalists. These groups have been partially and periodically successful in challenging the monopoly of evolutionary theory in primary and secondary education—legislators in at least ten states have considered mandating teaching evolution as just one theory and "intelligent design" as an alternative.[40] Yet this pressure has not cracked the orthodoxy of evolutionary theory in biology. There is no misbehaving science driving this controversy. It is essentially a political dispute separated from the scientific field.[41]

The social order of climate science, in contrast, is insufficient to allow scientists to quell disruptive public controversy and to establish their authority publicly. Climate scientists work under conditions of extraordinary scrutiny where their actions are evaluated by contradictory standards. This is one lesson of the "Climategate" e-mail hacking scandal.[42] Confidential communications among scientists revealed them to be strategizing about criticisms and data. Defenders claimed this was "just talk" and incriminating sounding words were simply labels for routine scientific procedures. Critics said the e-mails reveal a conspiracy to doctor data to inflate the perception of danger from climate change.

The pressure has led at least one climate scientist to engage in tactics similar to the skeptics. In February 2012 journalists published leaked confidential memos from the Heartland Institute, a climate change denial think tank, revealing sources of its funding and plans to discredit science taught to schoolchildren.[43] After speculation of the source of the documents, climate scientist Peter Gleick, citing his frustration with organized climate change denial, revealed that he had obtained at least some of them by deceiving Heartland with a false name. Some suggested that Gleick's sin was minor, no different from what an investigative journalist would do.[44] However, journalist Andrew Revkin argued that these deceptive tactics have undermined the credibility of Gleick and other scientists that was based on their capacity to make a distinction between the integrity of science and the deceptive practices of the deniers.[45] The point is that the intense political pressure climate scientists face has muddied the norms of appropriate behavior.

Anomie

This controversy is driven in part by anomie—the scientists lack appropriate norms and structures. Anomie here is certainly not the absence of norms. Indeed, there are abundant rules for scientific participation on policy panels like the Intergovernmental Panel on Climate Change, precisely designed to confront charges of politicization or cronyism.[46] Yet climate scientists have not been able to produce practices and structures that defend them from destructive scrutiny or that successfully preserve their authority under such conditions. This is not to blame climate scientists or to deny the role of powerful interests in producing the situation. Indeed, climate scientists' achievements under these conditions of scrutiny and structural weakness are remarkable. Yet the persistent, ungovernable controversy linked to rogue scientists and normative contradictions qualify climate science as an example of misbehaving science.

Misbehaving science may be an emerging problem in several areas of biomedical research. For example, in the nutrition field public controversy rages about dietary advice. There has long been agreement on the basic contours of what constitutes a healthy diet.[47] However, the "molecularization" of nutritional analysis—what Gyorgy Scrinis has called "nutritionism"—has helped open up vigorous debate about the virtues of different nutrients and their proportions (for example, healthy and unhealthy fats and how much carbohydrates versus protein) for weight loss and, to a lesser extent, health.[48] There is a huge industry driving diet information and products, and even it is dwarfed by the food industry. These forces coupled with the field's fluid boundaries promote a robust heterodoxy. Thus nutrition advice abounds in the public sphere, and the scientific field has little ability to impose their definitions (only a sliver of nutritional and health claims are regulated by federal authorities).

The nutrition science field is relatively powerless to hold the influence of the interested parties at bay and to set its own standards for research and communication; persistent, ungovernable controversy is one of the consequences.[49]

In other areas of biomedical research there are signs of growing anomie. One is the rise of efforts to craft explicit norms for situations that were previously ungoverned or governed implicitly. Many medical schools and scientific journals are seeking improved regulation, or at least disclosure, of doctors' and researchers' conflicts of interest.[50] Journals are also struggling to cope with the commodification of authorship with the rise of publication planning, ghostwriting, and contract research organizations.[51] There is also rising concern that negative research results are not being published, that funding sources significantly affect the outcome of published research, and also that the rate of retractions is on the rise.[52] Further, there is profound anxiety about disconnects between research and clinical practice and the reasons doctors seem not to follow clinical guidelines.[53] I take these to be signs of anomie: the halting transformation of a system of authority where norms based on professional honor are no longer adequate to the production of trustworthy knowledge. It is not clear that this situation is producing persistent, ungovernable controversy generally in biomedicine, but public distrust of this system (both its integrity and its claims to represent what is best for patients) contributes to an array of ungovernable local public controversies; for example, rising skepticism of vaccines and orthodox explanations of many ailments and the pursuit of alternative medical therapies.[54]

Thus the field of behavior genetics stands out in its scientific neighborhood for its character as misbehaving science. However, it is but one example in what seems to be an important trend in contemporary science. My strategy in this book is to study misbehaving science through the in-depth analysis of behavior genetics' development. In the conclusion I offer some general reflections on what this case might tell us about the broader trend.

Characterizing Behavior Genetics

The argument of this book is that behavior genetics' misbehavior—its affliction with ungovernable, politicized, scandalous controversy—is due to its social and cultural organization and position as a scientific field.

Behavior genetics is fragmented and poorly institutionalized; its members struggle to secure material and symbolic resources and often experience status anxiety linked to their scientific identity, yet the field enjoys tremendous public attention. This book is a historical account of the development of this condition. Here I give an overview of how the field appears today, the period since the mid-1990s race controversy with which the book began. *Behavior vs Neuro*

It is easier to see behavior genetics' situation through comparison to the "well behaved" field of neuroscience. Both fields were founded circa 1960 at a moment when universities were expanding and many new sciences were emerging. Both are interdisciplinary fields facing the inherent problems of drawing together scientists with diverse training and professional interests. And both had similarly broad intellectual ambitions at their founding: They sought to understand behavior (broadly conceived) mechanistically and naturalistically, to synthesize the expertise of diverse scientists, and to create a subject-matter-delimited field (rather than exploiting a narrow intellectual niche or a particular methodology, framework, or technological innovation).[55] Since both seek to link human behavior to biological processes, they would seem to have the same potential to provoke controversy about human behaviors and differences. But they diverge sharply as well: neuroscience has many more resources, is better integrated, and has avoided the kinds of controversies behavior genetics has suffered (and the publicity it has enjoyed). Some of neuroscience's growth and stability, relative to behavior genetics, must be explained by its technologies of brain imaging and its ties to medical practice. Although some of its growth and stability is due to its more robust field organization and its different relationship to scientific controversies.[56] Let's look more closely at the two fields, first at their integration and disciplinary organization.

The term *field* evokes the image of sports fields, battlefields, even agricultural fields—all relatively bounded spaces that are more or less unified internally by the common endeavor occurring within. Following this image, behavior genetics is not a very "fielded" field: it is not particularly coherent, integrated, or well bounded. But neither is it random chaos. It is a highly differentiated space where pockets of tightly integrated scientists relate to each other and identify with "behavior genetics" to very different degrees. Behavior genetics resembles less a soccer field and more an *archipelago*. It is a collection of different subfield "islands" that

Behavior genetics = bunch of integrated islands

are each occupied by scientists defined by different mixes of disciplinary background, technical practices, research subject (humans, mice, and such), and scientific questions. Different behavior geneticists—residents of various islands—have multiple, competing attachments: to local identities, interests, and practices; to their disciplinary "mainlands" (psychology, psychiatry, genetics, neuroscience, and such); and to "behavior genetics" as a whole.

Figure 1.1 is a map of the disciplinary territory of behavior genetics. The area labeled "behavior genetics" describes the islands whose members would recognize that label: psychological behavior geneticists (who usually call themselves simply "behavior geneticists"), psychiatric geneticists, molecular geneticists, and animal behavior geneticists. They are surrounded by other subfield islands who share a substantive interest in the genetics of behavior but who don't identify with behavior genetics or conceive of themselves as engaged in a common enterprise or competition with those from other islands. I've labeled this *implicit* field the "genetic inheritance of behavior." It includes subfields like neurogenetics, evolutionary psychology, sociobiology, behavioral ecology, psychopharmacology, and various arms of developmental, evolutionary, and even agricultural biology. The archipelago of subfield islands is situated in a bay where the mainland is divided into different disciplinary territories.

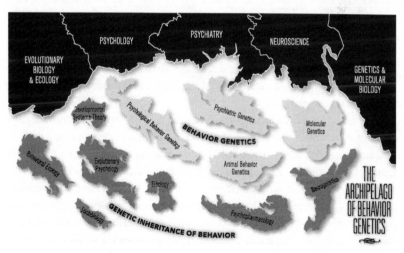

FIGURE 1.1. The archipelago of behavior genetics

Fragmentation along disciplinary lines is the rule within behavior genetics and in the broader space. Even the behavior geneticists who recognize themselves as engaged in a cross-disciplinary exchange are also strongly oriented toward their disciplinary homelands. This means they have multiple attachments—dual citizenship, to continue the territorial metaphor. Behavior geneticists seek recognition from each other, but even more so from their home disciplines.[57] These dual attachments can sometimes produce conflicts for behavior geneticists. For example, a long-standing dispute in the field is whether it is more valuable for research to appeal to concerns of neighboring disciplines or whether developing an autonomous research agenda would be more fruitful. Further, dual membership has allowed some behavior geneticists to disengage and retreat from major controversies rather than compelling them to engage in mutual policing and criticism. As we will see, the field's fragmentation, anomie, and persistent controversy are closely interrelated.

This fragmentation is reflected in the field's institutions. There are no "big tent" journals or societies that gather together behavior geneticists; discipline-specific venues are more important to them. Psychologist behavior geneticists studying humans dominate the Behavior Genetics Association and *Behavior Genetics*. At their founding circa 1970, however, they were conceived of as big tents. The aspiration of behavior genetics' founders was that the field would cover and integrate the whole implicit territory I've called the "genetic inheritance of behavior." The development of behavior genetics has been, in part, a history of sloughing off these different approaches. The history and ambiguity however leads some writers, especially popular ones, to use the label "behavior genetics" to represent the larger set of islands. This synecdoche offers behavior geneticists some of the authority of those far-flung scientists without any actual engagement with them.

In comparison, neuroscientists, though interdisciplinary, do conceive of themselves as part of a common field. Certainly there are many different subfields and approaches within—cognitive neuroscience, social neuroscience, molecular neuroscience, neuroimaging, and neurophysiology are but a few.[58] Some of these different approaches work under dramatically different experimental conditions and theoretical paradigms. A five-minute conversation with any neuroscientist will reveal that they can be passionately opposed to those in other subfields. Yet the difference is that they have an "imagined community," to use Benedict Anderson's term; they recognize each other as engaged in a com-

Common goal.

Behavior doesn't.

mon endeavor despite their disagreements.[59] A social neuroscientist and a neurophysiologist might have little in common scientifically, but both would accept belonging to the same field. However, among scientists fitting a substantive definition of "behavior genetics," a neurogeneticist, an evolutionary behavioral ecologist, and a psychological behavioral geneticist might hardly be aware of each other, let alone see some common purpose. Crucially, neuroscience's integration is encouraged by the Society of Neuroscience, which is a huge scientific society that provides a venue and representation for the diverse perspectives.

integration.

Let us compare the two fields in terms of their size, university institutionalization, and funding. First, size: Behavior geneticists number somewhat less than 1,200 scientists—the approximate sum of the membership of the field's three major societies.[60] In contrast, the Society of Neuroscience had over 42,000 members in 2011.[61] There are about ten journals that concentrate on behavior genetics, though its interdisciplinary research appears in many places.[62] However, there are over three hundred neuroscience journals, and the field has its own section in the Institute for Scientific Information's citation database.[63] In terms of overall growth, sociologist Joseph Spear has shown nearly exponential increases in neuroscience across many indicators. Growth in behavior genetics, measured in terms of the number of NIH grants, was in slow decline from the 1970s to the mid-1990s when advances in molecular and genetic manipulation techniques drew people and money to the field.[64]

Neuro ↑ #'s

↑ Neuro journals

Neuro e^x ↑

Behave ↓

The two diverge similarly in their disciplinary institutionalization within the university. Theorist Stephen Turner has argued that disciplines, conceived as labor market cartels for the production and consumption of graduate students, are an important way for scientific communities to regularize their own reproduction and offload the costs onto universities.[65] Neuroscience has a powerful disciplinary structure of over two hundred dedicated departments in the United States, in addition to being an interdiscipline that draws members from many fields.[66] Behavior genetics, in contrast, is an interdiscipline without a disciplinary structure. Behavior geneticists are educated and employed in other disciplines (psychology, genetics, psychiatry, neuroscience, biology, statistics, and so on). The market for behavior geneticists has never been very broad, and they have been highly concentrated in a handful of departments and institutes.[67] The lack of a disciplinary structure has also meant the "credentials" of members come from other fields, and the standards of expertise are all informal. The barriers to entry are thus few, though personal connections can be important given the informality of the field's structure and the concentration of its membership.

The fields show a similar pattern of divergence in the funding they receive. During the period of its most generous NIH support, from 2002 to 2005, behavior genetics received about $34 million in new grants on average annually; neuroscience, in contrast, received about $4 billion from NIH in 2005.[68] Neuroscience has about thirty-five times more members but receives roughly 120 times the funding. Information on private science funding is difficult to obtain, but evidence suggests that behavior genetics receives little, and most of that goes to psychiatric geneticists.[69] Neuroscience, in contrast, has many medical, pharmaceutical, and technological applications (for example, brain imaging), and has many ties to private, especially corporate, support. For example, pharmaceutical companies support about a third of the Society for Neuroscience's major prizes, and foundations fund most of the rest.[70]

There has always been a strong assumption that politically conservative organizations that are interested in propagating the idea that social inequality is genetically determined and therefore inevitable in a free society are an important source of funding for behavior genetics. This suspicion seems to be true in only a few cases. Race researchers Jensen, Rushton, Richard Lynn, Linda Gottfredson, and others, have long been backed by the Pioneer Fund, whose support of scientific racism and segregationist politics is well documented.[71] Murray and Herrnstein's *Bell Curve* was supported by the right wing Bradley Foundation.[72] And the famous Minnesota Study of Twins Reared Apart received funding from Pioneer and several rich individuals, including libertarian billionaire David Koch, as well as more traditional research funders like NIH, NSF, and the Spencer Foundation.[73] Conservative money has funded some of the most provocative and outspoken of the field's members, but they are the exception. The majority of behavior genetics projects and researchers do not receive such support.[74]

In contrast to its fragmentation, weak institutionalization, limited support, and especially its small size, behavior genetics has received a tremendous amount of public attention. For decades its claims have appeared regularly in major media sources, sometimes several per month. Some of this prominent exposure has been due to the heated controversies about particular claims. For example, coverage was intense for episodes like the IQ and race differences controversies following Arthur

Jensen's work in 1969 and *The Bell Curve* in 1994, Dean Hamer's "gay gene" claims in 1993, and claims throughout the period linking criminality to genetic or chromosomal abnormalities.[75] Behaviors from gambling to religiosity, to divorce, to financial choices, to popularity, to alcohol use have been staples of science reporting.[76] Further, in coverage on general topics like education, parenting, crime rates, or health behaviors, it is common for behavior genetics to make a cameo appearance as an aside about how scientists believe the phenomenon under discussion is under partial genetic control.

Prominent geneticists have often invoked behavior genetics in their public pleas for support and attention. In promising to unlock the "secrets of human nature" and to read from the "book of life," geneticists may have in mind working out the biochemical pathways that regulate DNA transcription, but they capitalize on the public's desire to know what makes some people smart or dull, good or evil. Dorothy Nelkin and Susan Lindee have shown how DNA has become a protean cultural resource for telling narratives about human differences and potential.[77] Behavior genetics has thus been a powerful symbolic tool for stoking the public's hereditarian curiosity. Behavior geneticists seeking public attention and other geneticists interested in drumming up support for the Human Genome Project have found value in appealing to this curiosity.

While the amount of publicity behavior genetics garnered is certainly out of proportion with its size, funding, or (as we will see) the degree of scientific respect it commands, this attention has not been matched by the institutionalization of behavior geneticists' expertise. Although the "policy implications" of behavior genetics have been extensively debated in the abstract, behavior geneticists have not been active participants in policy design or implementation.[78] According to behavior geneticist David Rowe, part of this exclusion is due to the concern of policy makers and educationists not with the origins of traits, but with whether this or that intervention works.[79] Another part is due to behavior geneticists' own reticence because of "abuses" of genetics in the eugenic past and the angry responses such efforts have drawn in the present.[80] Observers have noted that instead, behavior genetics arguments have appeared mostly as post hoc justifications for policy positions—usually regarding disinvesting in education or social support programs.[81] Behavior geneticists may have been enthusiastic about *The Bell Curve* in part for the chance that it could bring them greater attention and interaction with the policy world. As historian John Carson has shown, the inventors of intelligence tests in the early twentieth century promoted a social technology that remade crucial American institutions and opened up a debate about the nature of American democracy.[82] In contrast, the behavior geneticists, many of whom are their intellectual heirs, have energized a broad, but always contested, cultural conversation that has had more to do with individual responsibility and destiny than democracy—with limited institutional impact.

It is helpful to discuss the above in Bourdieu's terms of the volume and composition of capital—the material and symbolic resources deriving from different social realms (economy, politics, science, among others) that endow status and enable action. The overall volume of behavior genetics' capital is low. In material terms it is a small field with a weak infrastructure and limited resources, especially compared to neuroscience. The symbolic aspects are weak too, though this will become clearer in later chapters. Its controversiality is testament to behavior geneticists' shaky scientific authority. What about the composition of capital? Behavior genetics is relatively autonomous from the biomedical/industrial complex, and it lacks concrete ties to the world of policy and politics. Behavior genetics is richest in scientific capital. Yet its members can do a strong side business in media capital. Thus its relatively low volume of capital is a mix of these two types, which, as we will see, are in tension with each other.

What are the overall boundaries of behavior genetics? Steven Epstein's *Impure Science* showed how the field of AIDS research had highly permeable boundaries such that in addition to research scientists, public health officials, politicians, journalists, and especially social movement activists were all participants struggling to make legitimate claims affecting the science.[83] Neuroscience combines academic research scientists, clinicians, agents of pharmaceutical companies, and medical device manufacturers, among others. Behavior genetics is not like this—overwhelmingly, its actors are scientists who act in the domain of academic science. To a surprising degree behavior genetics is a "pure science," though it challenges what we might mean by that.

Behavior genetics is thus a paradoxical field. It is both coherent and incoherent. We will see broad disagreement on what it is, or even whether it is a field. It is scientifically prominent but with status and resource problems. It is culturally notorious but minimally institutionalized. The next six chapters explain how behavior genetics got this way as it encountered and coped with different kinds of controversies. The first of these next chapters considers the 1950s and 1960s when, well aware of the dangerous legacy of eugenics and scientific racism, the field's founders sought to establish behavior genetics on terms that would immunize it against controversy.

CHAPTER TWO

Founding the Field to Avoid Controversy

In the first few decades of the twentieth century, the field of genetics was intimately linked to eugenics. For most scientists, interest in dissecting the patterns and mechanisms of genetic inheritance was motivated by the desire to improve the gene pool, to proliferate the genes associated with socially desirable traits and restrict or eliminate those linked to negative traits. The traits of greatest concern, feeblemindedness, imbecility, pauperism, criminality, prostitution, alcoholism, and the like, were the vernacular social problems of the day. To many privileged social groups the seeming concentration of these traits among the growing laboring classes, immigrants, darker races, and inhabitants of the urban ghettos represented a looming danger. For many elites, such traits—as well as their moral obverses, eminence and genius—had long been considered organic, immutable, and hereditary. The emerging science of genetics promised to rationalize these sentiments. Genetics could unlock the causes or at least the rules of transmission of these traits and offer tools to manipulate them.[1]

Although never without controversy, eugenics enjoyed wide support among scientists and much of the public until World War II. Scientists launched massive data collection efforts among military recruits, at immigration centers, in schools and asylums and assembled family pedigrees to map the incidence of particular traits among different classes and ethnic and racial groups. Ordinary people participated in "fitter family" contests to represent eugenic ideals. Heeding eugenicists' advice, many states enacted laws to enable sterilization in prisons and insane asylums, and the federal government severely restricted immigration in 1924. And then, as is well known, inspired by the great progress in America, the Nazis enacted eugenic policies on a massive scale, starting with the sterilization of the disabled and the insane and ending with the Holocaust.

The revelation of Nazi atrocities and the more democratic and egalitarian mood of postwar America undermined the broad enthusiasm for eugenics. Eugenics and the political involvement of their science more generally, threatened to discredit geneticists and associate them with the sorts of racism and antidemocratic politics against which Americans had just gone to war. Although it never disappeared, geneticists' interest in eugenics became less widespread, less vocal, less entangled with their research, and, when expressed, much more circumspect and moderate. If many still believed in hereditarianism—that behavioral traits are largely genetically determined—they nevertheless tended to pursue research that sidelined that issue. Many abandoned the search for differences among classes, races, and ethnic groups, turned their attention away from social problem traits toward uncontroversial medical problems, and became absorbed by molecular mechanisms of heredity.[2]

Eugenics and hereditarianism were not the only fraught issues here, so too were the conceptualization and measurement of behavior and intellectual capacity. The eugenics movement was not interested in "behavior" strictly speaking. It was concerned with human character and quality, *who* people were, not how they acted. The question of character was tied up in political concerns about the organization of society. As historian John Carson has described, during the first quarter of the twentieth century psychologists, particularly those developing intelligence tests, were focused on providing a technical solution to the contradiction between equality and merit in American democracy.[3] Equal treatment and opportunity are central to the ideology of democracy, but there are clear inequalities of capacity and wealth among people. Are these due to different inherent abilities or the unfair distribution of unearned resources? Psychometricians aimed to develop tests able to identify intellectual merit objectively (as well as the demerit of feeblemindedness, criminality, and the like). Such efforts touched the science of heredity because they aimed to identify *innate* ability, or that part of inequality they argued wasn't unfair. Beyond heredity, these psychologists aimed to provide a scientific definition of behavioral merit, the means to measure it, and to establish themselves as authorities. Carson shows that the mental testers' efforts were tremendously controversial. At stake was "who should have the power to define what was and was not equal, democratic, and fair."[4]

At midcentury this tangled legacy discouraged those interested in the genetic inheritance of behavior from actively organizing. The ascendance of behaviorism as an intellectual movement had already moved psychology away from the concept of character and toward a focus on behavior—actions in response to stimulus. But behaviorism's focus was often on the situation, it did not ask about the origins of response to stimulus or how these patterns of behavior were inherited. Therefore, the effect of genes on the behavioral nexus was an obviously important and untapped intellectual question to many geneticists, psychologists, anthropologists, sociologists, psychiatrists, zoologists, agricultural researchers, and physiologists. It offered the possibility of bridging interests in the social, behavioral, and biological sciences. More than an intellectual opportunity, it offered the chance to capitalize on the postwar expansion of science in an era when new interdisciplinary fields (neuroscience, cybernetics, and medical genetics, for example) were rapidly forming in this space.

In the past, researchers had been unable to approach the question without the usual presumption being that behavior is determined by heredity and that the point was the measure of social merit, the ranking and sorting of human quality. Could this framework be broken? Was it possible to build a genuine science of behavior genetics without being consumed by the political and scientific problems of hard hereditarianism, racist eugenics, and the merit paradigm?

The possibility of forming behavior genetics on scientific, not politicized grounds was the field's first great controversy. It was an implicit controversy—problems the field's founders were trying to avoid. This chapter shows how they coped with the specter of controversy and how it conditioned the way they organized the field socially and intellectually. This period before the IQ and race controversy of the 1970s was the field's high-water mark for intellectual breadth, diversity, and integration. In many ways it was the field's golden era.

It would be inaccurate to interpret early behavior geneticists' efforts to head off this problem as strategic, at least not in two common senses. They were not trying to conceal an actual commitment to eugenics or racism under a veneer of "science." Nor were they engaged in marketing to downplay the questionable elements of the new science or its social implications to make it acceptable. They would, however, be accused of doing both in future periods of controversy. Rather than concealment, they were quite interested in the social implications of the science and what its relationships to eugenics might be. Their objections to eugenics and racial comparisons rested less on political or ethical objections than technical scientific criticisms. They believed, for example, eugenic policy and racial comparisons rested on misunderstandings of genetic variation at the individual level and the environmental dependence of gene action. Further, they worried whether socially interesting topics like intelligence and criminality could be characterized objectively, and they believed others—like perceptions, reflexes, and language—might be more scientifically tractable. Thus the early behavior geneticists were fully committed to this new field and to constructing it on its own authentically scientific terms that would differentiate the field from its ignoble forebears. "It is not a coincidence that genetics has been the biological science most prostituted in both Fascist and Communist states," warned a 1960 textbook.[5] A 1967 textbook stated that the aim, therefore, was "to discuss how it is possible to obtain *reliable* and *unbiased* statements on the relative influence of heredity and environment in the control of behaviour."[6]

There were social and intellectual dimensions of this founding project. The social story focuses on early behavior geneticists' efforts to form a network of interested scholars. In the 1960s behavior geneticists obtained grants and institutional support for building their network. They actively recruited scholars from several different disciplinary backgrounds working in a variety of paradigms. But their efforts were selective. They kept a degree of distance from several figures who had emphasized eugenic or race-comparison themes in their work. They pursued support from institutions that didn't make political demands on their work. And they associated closely with Theodosius Dobzhansky, who was one of the most visible geneticists of his day and a staunch public opponent of racism and eugenics. My account of this era contradicts those critics and observers who have tended to see behavior genetics as the direct progeny of a dark hereditarian past.[7]

There are other surprising findings from this era. One is that the young field was founded and led mostly by students of *animal* behavior. This is striking in two ways. First, it is the opposite of the earlier eugenic era when those studying *humans* dominated claims about the heredity of behavior. Second, it is very different from the mid-1970s after which behavior genetics became nearly synonymous with human research, and animal studies faded from the field's broader image. A widely held sentiment among behavior geneticists (human and animal researchers alike) was that animal research would be crucial to leading the field in a non-controversial direction. They hoped animal research could be a model for and complement to human research that would help orient the field toward academic, not applied, questions. The aim was not to suppress the psychometric tradition, with its ties to eugenics and the merit paradigm, but to improve it through interdisciplinary exchanges.

The field's interdisciplinary agenda was broadly appealing during the 1960s in another surprising way. Now viewed as the site of trench warfare between scholars from inherently opposed intellectual traditions, at this point they saw much common cause.[8] I will show that some of the population geneticists who would become behavior genetics' most mortal enemies in the IQ debate were, in the mid-1960s, happy participants in the field's construction. Other intellectual paradigms that would later become competitors to behavior genetics—developmental genetics, social psychology, comparative anthropology, behavioral ecology, and ethology, for example—were, at this time, eagerly courted.

In the end, I argue, behavior geneticists sought to avoid the potential controversy stalking the field through an implicit pact among the membership. Diverse researchers would collaborate and support each other's work, lending each other scientific tools and legitimacy. Hereditarian psychometricians who were closest to the earlier, problematic tradition, would be only a minority among the animal researchers and others. The condition of this cooperation would be an agreement that the field was not about reviving eugenics or racial research. The intellectual concomitant of this pact—beyond avoiding politicized purposes and topics—was a broad and diverse scientific agenda.

An agreement about what this science was about

Assembling the Field

Conferences, Organizations, and Networks

As historians have pointed out, the Jackson Laboratory in Bar Harbor, Maine, was a crucial node in the early organization of the network that would become behavior genetics.[9] In the 1930s Jackson Laboratory had established itself as an important center of research in mammalian genetics and cancer genetics, and it would eventually become the main US supplier of lines of inbred mice crucial to genetics research. In the early 1940s Alan Gregg, a program director at the Rockefeller Foundation, agreed to fund a ten-year program at Jackson Laboratory on the genetics of behavior in dogs. One aim was to breed a highly intelligent pet dog that would be commercially viable and help convince the public of the high heritability of intelligence.[10] John Paul Scott, who later described himself as the only trained geneticist who at the time was interested in behavior, was recruited to run this Division of Behavioral Studies in 1945.[11]

Scott quickly organized a pioneering conference on "Genetics and Social Behavior" in 1946.[12] The conference assembled scholars into committees on topics like "genetic background of behavior," "intelligence and learning," and "social organization and leadership." It introduced attendees to the facilities and emerging projects at Jackson Laboratory and helped establish the lab as a key research and training center. Over the next twenty years many scientists did stints at Jackson Laboratory working on its projects or attending its short courses and training programs. These projects and researchers produced dozens of papers, mostly concerning dogs and mice, that would form much of the empirical base of the emerging field.[13] In 1965 Scott and John Fuller, his research partner, published *Genetics and the Social Behavior of the Dog*. It was the most important publication to come out of the Jackson Lab program. Scott and Fuller found many genetic differences between dog breeds; they did not find evidence for general intelligence or temperament. Dogs would exhibit different degrees of intelligence or temperamental characteristics depending on the situation. This evidence of interaction led them to question the high heritability of human intelligence—thus undermining a goal of the Rockefeller Foundation sponsors who had hoped to discredit the idea that intelligence was the product of education.[14] Although the behavioral program at Jackson Laboratory declined after this point, it had been the first important base for the new field. —Jackson Lab.

In the 1950s there were different ideas about what to call the emerging field. Scott debated calling the field "biosociology" or "sociobiology," choosing the latter "on the analogy with biochemistry; the more basic science coming last."[15] In 1951 Calvin S. Hall, another associate, wrote a review chapter and labeled the field "psychogenetics," presumably on a similar principle to Scott.[16] But in 1960 Fuller and William Robert Thompson, an animal psychologist, wrote the field's first textbook, *Behavior Genetics*. This was the name that stuck, and behavior geneticists have long marked this publication as the field's founding document. There is no suggestion of conflict over the name at the time, but the possibilities do reflect slightly different visions. Scott's label empha-

1960 Founding textbook

sized the social dimension of behavior—a framing that has since been
weak in the research.[17] Hall emphasized the combination of the disciplines of psychology and genetics, which was perhaps narrower than the
interdisciplinary network being assembled. Fuller and Thompson chose
behavior genetics to find "the most general and acceptable name" that
would appeal beyond psychology while also avoiding a term that would
cause conflict with "genetic psychology" (more commonly called developmental psychology today), "which has at times been rather antagonistic to heredity."[18]

In the 1950s, meetings, seminars, and short courses began to appear
in various locations, spreading the behavior genetics network. In 1954
the American Psychological Association held a symposium on behavior genetics.[19] And in 1959 the American Psychiatric Association held a
major symposium at the American Association for the Advancement of
Science meeting called "Roots of Behavior," which produced a volume
on animal behavior genetics.[20]

While these events raised the profile of the young field, even more
important for strengthening the network of participants was a pair of
three-week-long meetings held in the summers of 1961 and 1962 at the
Center for Advanced Study in the Behavioral Sciences in Stanford, California, and sponsored by the National Science Foundation.[21] Organized
by fruit fly researcher Jerry Hirsch and mouse researchers Benson Ginsburg and Gerald McClearn, these meetings gathered more than twenty
researchers from across the country and locally. This effort spawned a
summer training institute at UC Berkeley supported by the Social Science Research Council (SSRC) and the National Institute of Mental
Health (NIMH) and a volume of long, detailed topical articles. Many of
the participants in this meeting—Hirsch, McClearn, Ginsburg, Thompson, mouse behaviorists Jan Bruell and Gardner Lindzey, developmental
geneticist Ernst Caspari, and psychiatric researcher Lissy Erlenmeyer-
Kimling to name a few—became central figures in the early organization
of behavior genetics.

Another set of conferences sponsored by the American Eugenics Society (AES) held at Princeton in the 1960s led to the establishment of
the Behavior Genetics Association. Starting in the late 1930s, under
the leadership of Frederick Osborn, the AES had begun to reorganize
and "reform" eugenics, separating it from political movements, shifting the concern from ideal biological types of people toward recognition
of human variation, and connecting the sciences of biology and demog-

raphy.[22] From 1953 to 1971, the AES sponsored symposia and conferences on topics like intelligence, genetic counseling, and fertility trends and published the proceedings in its new journal *Eugenics Quarterly*.[23] In 1964 the AES brought together demographers and population geneticists in the First Princeton Conference. In the next two years psychologists, anthropologists, sociologists, and animal behaviorists were included as interest grew in linking demography, genetics, and behavioral science. At the Third Conference, Dobzhansky proposed forming a permanent scientific society. Osborn pledged modest financial support, and over the next three years a somewhat byzantine process involving surveys, meetings, and committees resulted in the Behavior Genetics Association's establishment in 1970. The BGA would hold annual meetings and take on the journal *Behavior Genetics*, which had just been established independently.[24]

Patterns of Participation

The establishment of the field's society and journal mark the end of behavior genetics' first period. It is worth noting some of the patterns of participation that characterized it. First, it was animal behaviorists who took much of the lead in building the behavior genetics network— *Not supporters of eugenics* through synthetic writing and research, organizing events, and training others. Participants in the emerging field came from a variety of disciplinary backgrounds. Psychology and genetics were represented prominently but certainly not exclusively. Also participating were anthropologists, statisticians, agricultural scientists, physiologists, medical researchers, anatomists, and primatologists. Psychometricians whose forebears had focused on the heredity of behavior with their claims about feeblemindedness and racial degeneration were not the architects of the field.[25]

Participants did not envision behavior genetics as a disciplinary subfield. Most participants wanted the field to be more than a combination of the disciplines of psychology and genetics. Although there was sometimes a tendency to describe the field in such terms—the editors of the inaugural issue of *Behavior Genetics* described it as "simply the intersection between genetics and the behavioral sciences"—there was no boundary work to exclude disciplinary perspectives.[26] Overall, participants saw it as a multidisciplinary space that would be open to scholars interested in the heredity of behavior from whatever direction. Indeed, when there were disciplinary conflicts, for example in the Second

Princeton Conference when geneticists and psychologists came into conflict over the conceptualization of variation (a matter I consider further below), participants viewed the solution as more engagement, not bounding anyone out of the field.[27]

In these early days, scholars who would during future controversies face off as enemies cooperated in the forming of behavior genetics. Participants in five Princeton Conferences included population geneticists Richard Lewontin, Luca Cavalli-Sforza, and Walter Bodmer along with many behavior geneticists, including those from psychology like Irving Gottesman, John Loehlin, and Sandra Scarr. A few years later, the former group would harshly attack behavior geneticists for their handling of claims about genetic causes of racial intelligence differences. This story is told in depth in the next chapter, but it is worth noting that prior to this controversy, there was much goodwill about the field, diverse scientists were interested in its potential, and now taken-for-granted disciplinary divisions then did not exist. The connections at this moment in the field's development upend the view that professional or epistemological differences make conflicts over behavior genetics inevitable.[28]

Associations and Dissociations

While behavior geneticists were open regarding disciplinary participation in the field, they were cautious about associating with particular figures who had been tied to the eugenics movement or took extreme positions on the genetic determination of behavior. Had the field's organizers sought to include all prominent biological and behavioral scientists interested in the heredity of behavior we might have expected contemporary advocates of racial differences or eugenics to be involved. Prominent figures who at the time had made strong arguments about race—like psychologists Henry E. Garrett and F. G. McGurk, physiologist Dwight Ingle, geneticists C. D. Darlington and K. Mather, and anthropologist Carleton S. Coon—did not appear in the network.[29]

Some exceptions are worth noting. Robert M. Yerkes, the pioneering intelligence tester who warned that new immigrants and blacks threatened "race deterioration," had been made the chairman of the 1946 conference at Jackson Laboratory because of his status as the unofficial "dean of comparative psychologists."[30] Raymond Cattell, the eminent psychometrician who had pioneered several important statistical techniques, but had also written admiringly of the eugenic potential of fas-

cist social policies, had played a role in organizing a meeting in Louis- *Exceptions*
ville, Kentucky, that would culminate in the 1965 volume *Methods and* *to assoc.*
Goals in Human Behavior Genetics.[31] And the patron of "reform eugenics," Frederick Osborn, funded the Princeton Conferences and sup- *w/ Eugenics*
ported the formation of the Behavior Genetics Association.[32] However,
the influence of these figures seems to have been focused on these particular events. They did not appear in other venues, nor are they credited
as active in behavior geneticists' accounts at the time.

Several factors explain the potentially provocative figures' limited influence. Some of them were sequestered across the Atlantic, and while
behavior geneticists imagined their field as international, they focused
their efforts on organizing in the United States. Further, the most active organizers of behavior genetics were younger scholars at the beginning or the early-middle part of their careers. It wasn't just older figures
who were tied to eugenics who didn't participate; neither did those elders, like psychologists Robert C. Tryon and Edward C. Tolman, whose
maze-running experiments involving selectively bred rats were influential on behavior geneticists. Behavior geneticists were establishing a new
science that would not be beholden to a generation of elders.

However, some of the avoidance of controversial figures was clearly
intentional. For example, a letter between members of the SSRC Committee on Genetics and Behavior concerning whom to invite into the
group included the lines, "The other names mentioned in your letter are
mostly just names—some of them very distinguished ones—from the literature to me. I think it might be wise to go slowly on Kallmann because
of the rather polemical nature of his work as well as his mixed status in
the field."[33] They were referring to Franz Kallmann, a German emigrant
psychiatrist at Columbia University whose views of the strong genetic
determination of mental disorders, and support of eugenics, many of his
contemporaries deemed extreme.[34] In this instance and others, members
were trying to avoid hereditarian hard-liners who might complicate their
efforts.

Behavior genetics had a similarly distant relationship to eugenics via
its institutional support. The Rockefeller Foundation officers who autho-
rized support for Scott and Fuller's study of the genetics of dogs' behavior had eugenic motivations. But the fact that the scientists' conclusions
stymied these expectations demonstrates that the new breed of behavior geneticists were serious about their scientific bona fides and autonomy. Historian Diane Paul claims that before the late 1940s those inter-

ested in the heredity of behavior were funded primarily by individuals and organizations eager to support those willing to make eugenic pronouncements. By the 1960s several things had changed. Governmental support for science had boomed. Foundations were supporting crucial experimental infrastructure, like expensive registries of twins, yet their motives were less about eugenic policy than influencing behavioral and social scientists to be less fully environmentalist in their assumptions—views fully compatible with behavior geneticists' own.[35]

The other side of these exclusions was the field's strong association with Theodosius Dobzhansky. Dobzhansky was a Russian émigré who worked at Columbia University, Rockefeller University, and UC Davis on population genetics using fruit flies. As a public intellectual, from the 1930s to the '70s he was one of the strongest voices for a nonracist, noneugenist, nondeterminist vision of genetics that would explore and value human genetic diversity.[36] A prolific mentor, he trained several behavior geneticists (and some eventual critics of the field). He served on the SSRC committee that helped coordinate the behavior genetics network in the 1960s. He is credited with suggesting the idea of forming the Behavior Genetics Association and served as its first elected president in 1972.[37] Dobzhansky had connections that helped secure visibility and support for the young field. But more importantly, behavior genetics' association with his public brand of progressive genetics signaled the kind of field members hoped it would be.

Leaving Behind a Shameful Legacy

As they were assembling this network and establishing its institutional supports, behavior geneticists were also articulating the field's intellectual commitments and scientific agenda. Ostensibly these were about understanding the heredity of behavior, but an abiding concern was avoiding the implicit controversy of building this science without reviving the specters of racism, eugenics, Social Darwinism, and crude biological determinism. They did this in two basic ways. First, rather than ignoring the issues of race and eugenics, they explicitly confronted them to show that both were politically and scientifically wrongheaded. Second, they sought to establish in positive terms what behavior genetics should be. They outlined a heterogeneous and interdisciplinary agenda, an impli-

(1) Confronted eugenics
(2) Establish behavior genetics, +

cation of which was that it would not be caught up in narrow, politicized interests.

Eugenics and the Science of Superiority

Behavior geneticists understood that politicization was an ever-present danger. While wanting to preserve the field's potential for application to social problems, they didn't want it to be appropriated for political projects or perceived as being politics merely disguised as science. Given the field's prehistory, as well as contemporaneous and recently past events, the two crucial arenas of politicization were eugenics and racism. Behavior geneticists knew that in order for it to be viable, they would have to distance the field from both. They saw this task mainly as scientific. The political excesses of eugenics and racism, they felt, were often dependent on misconceptions about technical matters. Almost never were their critiques strictly political, ethical, or moral—for example, that interfering in individuals' reproductive sovereignty was wrong or looking for racial differences would always be politically destructive. Yet they often announced broad prodemocratic, prosocial, and antiracist political values as part of their orientation.

Eugenics, racial differences, and politicization were not constant preoccupations, yet they did appear occasionally in the literature during these early years. Conferences and edited collections sometimes devoted space to these matters. A 1968 conference at Rockefeller University even featured specialists in this history.[38] It is noteworthy that if this engagement was not exactly enthusiastic, neither was it reluctant. Certainly behavior geneticists were trying to clear space for what they really wanted to do, and they were demonstrating their goodwill and responsibility. It was an important task. As psychologist Irving Gottesman noted, a "source of reluctance to become involved in behavior genetic research is the lurid and disquieting history of the eugenics idea."[39]

But more than that behavior geneticists seemed to think that important scientific issues could be elucidated through their discussion, and also that it was important for them and their students to be educated about these issues, perhaps to avoid the mistakes of the past. For example, a strong implication of eugenicists' concerns with genetic purity and the propagation of deleterious mutations was that genetic homozygosity—where both of the two copies of a gene we all carry are the same—would have to be privileged (in terms of fitness or whatever). Dobzhansky, a strong opponent of this view, argued that heterozygosity—where the two gene copies are different forms—was advantageous to individual creatures, and genetic diversity aided the fitness of populations because it gave them the genetic resources to be adaptable to changing environments.[40] The implication for behavior genetics was to abandon any search for superior or inferior genotypes because, even if possible to identify, these would be context specific. As Fuller and Thompson explained, "since we do not know the nature of the future society to which our descendants must be adapted," eugenic recommendations must be limited.[41]

Perhaps the archetypal anxiety of the eugenics era was the idea that "feebleminded" and "moron" individuals were greatly outbreeding their social "betters." This notion had been shown to be based on a biased estimation: very few low intelligence individuals had children, though the ones who did tended to have many. Their total fertility was actually less than those with high intelligence, and thus the eugenic anxiety was empirically groundless.[42] Historian Edmund Ramsden has shown how important this finding was to those hoping to connect the disciplines of genetics, behavioral science, and demography.[43] It allowed them to distance themselves from politicized eugenic claims on scientific grounds while keeping alive questions of population, genetics, and behavior at this scientific intersection. Field leader Jerry Hirsch discussed an updated eugenic anxiety, which was that "Great Society" efforts of education and meritocratic mobility would enhance assortative mating by intelligence and could "serve to further stratify society into a rigid caste system."[44] But Hirsch was reassuring that this was unlikely for two reasons: First, genetic recombination and imperfect inheritance would produce regression to the mean, where the offspring of those on the extremes of a distribution tend to be closer to the middle. This was an outcome Hirsch saw as positive but that Sir Francis Galton and other eugenicists bemoaned. Second, a complex society would open up new social niches "to be filled by each generation's freshly generated heterogeneity."[45] Thus behavior geneticists were eager to point out that the greatest eugenic fears (and hopes) were unlikely to come to fruition.

However, behavior geneticists did not dispute a basic premise of eugenics—that inherited behavior had social structural implications. In 1971 Irving Gottesman and Lissy Erlenmeyer-Kimling wrote "A Foundation for Informed Eugenics," which tried to articulate a version that valued "rational elitism and an open-class democratic society . . . [and] the dignity of mankind" while breaking with the "cranks, racists, bigots, some well-meaning souls, and miscellaneous zealots" who had long infected the movement.[46] They defined eugenics as a social luxury that should be pursued after "higher priorities . . . [such as] war, poverty, pollution, unequal civil rights, and hopelessness" had been eradicated and global population growth stabilized.[47] Future eugenics would focus on undisputed physical and mental disorders, abandon the "myth of ideal types" of humans, and develop an "Index of Social Value" rather than fixating on IQ maximization.[48] Thus behavior geneticists were not rejecting eugenics in principle but showing it to be simultaneously compatible with and dependent on liberal, "Great Society"–type programs and also nothing with which their current research was concerned. Why this research is more important now than eugenics

Racial Differences

Alongside eugenics, the issue of whether different races differed in their genetic potential was a crucial matter for behavior geneticists to confront. As with eugenics, behavior geneticists' efforts to control the issue of genetic racial differences were aided greatly by their association with Dobzhansky. Targeting both scientific and lay audiences, he had long argued against racial hierarchy as an implication of modern genetics.[49] He often repeated the slogan, "differences aren't deficits" and argued that equal legal treatment had no dependence on the biological equality or inequality of individuals or races. Part of Dobzhansky's argument against racial hierarchy was evolutionary—forces of natural selection were unlikely to have selected races to be behaviorally inferior; rather, they promoted behavioral flexibility and educability across humanity. With geneticist L. C. Dunn, Dobzhansky argued that "genetically conditioned educability has guaranteed the success of mankind as a biological species, and has, in turn, permitted progressively more advanced cultural developments."[50] They did not deny the possibility of racial genetic differences in "mentality" but say that they would necessarily have to be "not fundamental" to have survived through evolutionary history.

In their field-defining 1960 textbook, behavior geneticists John Fuller and William Robert Thompson echoed the argument: "It is difficult to conceive of a society in which intelligence, cooperation, and physical vigor would not have positive selective value. Hence it is likely that natural selection tends to oppose the establishment of major heritable differences between races. . . . Natural selection should effectively prevent the evolution of a race of morons, but it would not eliminate them from a society which had a need for unskilled workers."[51] Their point was that inferior races would not have survived evolutionary history, but that genetic differences among classes could have emerged as societies became more complex.

Another set of arguments concerned the ways that contemporary behavior geneticists thought about traits, populations, and individuality. Dobzhansky often repeated that "what is inherited is not this or that trait, but the manner in which the organism responds to the environment."[52] What this implied is that even if racial groups differed on average in the genetic "responses" they inherited, and these were related to important behaviors, they would change in different environments. Thus there is no such thing as genetic inequality net of environments.[53] Further, contemporary population genetics had undermined the traditional "typological" conception of races. Rather than seeing races as groups with different genes or alleles, they were understood as groups with different *frequencies* of the same genes and alleles.[54] All races would contain some individuals that had inherited all the different genetic responses to environment. This view of races as genetically overlapping dramatically restricted the ways it was possible to talk about them as genetically different and completely undermined any genetic information about particular individuals that could be gained by identifying their race. Finally, claims about genetic racial differences depended, in part, on the idea of races having typical or normal individuals. Behavior geneticist Jerry Hirsch argued that the incredible variability of the genome (including the many mechanisms for producing variation) makes each individual genetically unique. When combined with the different environments in which the individual might develop, there are unlimited combinatorial possibilities for heredity/environment interactions producing the individual's phenotype. Hirsch believed this radicalized the concept of individuality and destroyed the idea of a normal individual.[55] These ways of conceiving traits, populations, and individuals did not suggest that races were perfectly equal, but rather showed that the possibilities for talking about racial differences were limited and uninteresting compared to other ways of considering behavioral variation.

Behavior geneticists' concern with inbred strains of animals produced another line of argument about racial differences. If, following the geneticists' definition, races were simply populations that had different allele frequencies, then inbred strains of animals—mice, rats, and dogs in particular—could stand in as models of human races. In their textbook Fuller and Thompson write, "Strains of animals show behavioral differences correlated with their diversity in genotypes, and it can be argued that the same must be true of human races."[56] However, several other notions qualified this point. For one, these animal "races" had been artificially selected to be more homogenous within and more different between, so their differences would likely be exaggerated relative to those among human races. Further, many animal studies showed that strain differences in behavior were not independent of environment. R. M. Cooper and J. P. Zubek's study of rats selectively bred to be "dull" and "bright" in maze-running ability showed dramatic differences between the strains in the "normal" environment. But in the "enriched" and especially the "restricted" developmental environments, both strains' performance were quite similar.[57] Scott and Fuller made a similar finding in their comparative study of dog breeds: "The behavior traits do not appear to be preorganized by heredity. Rather a dog inherits a number of abilities which can be organized in different ways to meet different situations."[58] Thus even creatures that had been explicitly engineered to embody racial superiority and inferiority could not demonstrate the idea in any simple way. These findings helped behavior geneticists conclude that the racial analogy to animal strain differences "need not imply racial superiority, merely racial differences."[59] *No inferior/superior!*

Finally, in a volume of essays reporting the results of a set of meetings important to the establishment of behavior genetics, anthropologist James Spuhler and psychologist Gardner Lindzey reviewed the empirical evidence for racial differences in behavior. Beginning by asserting that "dictators, demagogues, and elitists" had misused race, they argued that this alone is not a good reason to abandon it.[60] But the evidence they found for genetic racial behavioral differences was limited to color blindness and capacity to taste the bitterness of phenylthiourea (PTU)—traits under simple genetic control whose allele frequencies vary among racial populations.[61] Although intelligence and personality tests show different racial distributions, they could find no good evidence that these differences were independent of environmental variation or the testing situation.[62] They concluded, "For the areas of human behavior that are vital in everyday life, for the varieties of behavior that allow individuals to participate satisfactorily in their society, there is no comparable evidence [to the color and taste perception results] for genetically

Can't separate race from environment, can't prove inferiority/superiority

determined racial differences."[63] They also argued that to study racial differences seriously, researchers should look at "simple processes and modes of response" more likely to be physiologically linked and less entangled in culture and language, but that research designs for demonstrating these differences unequivocally are not available in humans.[64] They state that overall "the concept of race is likely to remain of small general importance for behavioral science. . . . The possibility of future change in the status of the concept is dependent on increased activity in an area of research that is procedurally difficult, politically dangerous, and personally repugnant to most psychologists, sociologists, and anthropologists."[65]

The overall thrust of these efforts was to push the search for racial differences off the field's agenda. Without denying that such differences were a possibility, they gave evolutionary, methodological, comparative, and empirical reasons to doubt their magnitude, social importance, and scientific tractability. If there was debate, it was not about the evidence for racial differences or the likelihood that differences would be large. Some, like Lindzey, seemed to believe that in a world without politics and with perfect scientific freedom, race differences research would be desirable (if not a priority), while others, like Hirsch, viewed it as a complete pseudo-problem.[66]

As we have seen here, behavior geneticists were deeply concerned to keep their field from being consumed in the old controversies characteristic of previous efforts to link genetics and behavior. Historians of genetics have come to argue that race difference research and eugenics inextricably animated these early efforts.[67] Behavior geneticists, in contrast, saw them as politicizing contaminants that could be separated from the science. They sought to distance their field from race and eugenics not by suppressing or ignoring the topics but by engaging them. They undertook this task actively, willingly, and under their own initiative. This is noteworthy because in future years, especially in the throws of active public controversies, typically outsiders would bring up such concerns and behavior geneticists engaged them only when forced.

Behavior geneticists partly sought to clear space for the new field, but they also believed that there were matters of scientific importance that engaging race and eugenics could elucidate. For example, in disput-

Main arguments

ing the basis of eugenics they were explaining that reactions to environments, not traits, were genetically inherited and behavior was complex and malleable. In disputing the race concept they were articulating the field's nontypological understanding of populations and the central focus on individual (rather than group) differences. While behavior geneticists acknowledged there were political and ethical objections to race and eugenics, their own were mostly scientific and pragmatic. They thought the eugenics movement and much research on racial differences were plagued by empirical, conceptual, and methodological problems. These areas shouldn't be pursued because the scientific difficulties relative to the intellectual payoff were not worth it—especially given the context of political and ethical sensitivities and the open field of viable research problems before them.

It is noteworthy, finally, that there was essentially no dissent from these views in the field. No behavior geneticist argued that, yes, informing a eugenic program should be an aim of behavior genetics. None warned about the problem of gene pool decline, epidemics of hereditary imbecility, or the like. Nobody in the field argued that racial behavioral differences were actually scientifically tractable and important to study. For some, the reasons not to study racial differences were more political than scientific.[68] But this was not to argue that there were compelling scientific reasons to study racial differences. Thus there was a collective agreement that scientific boundaries against these topics had to be drawn, but this took place in the absence of a particular controversy or significant disagreement among behavior geneticists.

Defining the Positive Intellectual Agenda

Behavior geneticists had thus moved to decisively distinguish their field from genetic determinism, eugenics, racial research, and politicized framings of the field in general. At the same time, they carved out a positive vision of the field that was broad, ambitious, and cut across disciplinary lines. They tried to maintain the field's scientific and academic focus. The implications of the research for social policy were kept largely on the horizon. Instead, their main targets were intellectual problems in the academic community—battling behaviorist and environmentalist paradigms in the social sciences, encouraging focus on between-individual variation, and creating hybrid biosocial explanations of behavior. Ani-

Main targets

mal research dominated the field, and researchers, including many studying humans, hoped that its "pure science" orientation would inform the
approach to human research. A crucial debate was how much importance should be given to the nature versus nurture framing of research—
that is, quantitatively estimating heritability, which is the proportion of
the variance of a trait that's due to genetic (as opposed to environmental) variance. Many researchers were enthusiastic about the emergence
of new, high quality sources of twin and adoptee data to enable heritability estimates. But others saw heritability as a limited concept empirically
and conceptually. As a result behavior geneticists proposed a range of
ambitious, transdisciplinary projects and questions with the aim of leaving the nature versus nurture issue behind. Overall, behavior geneticists
were working to produce a definition of the field's scientific capital that
certainly combined scientific and political thinking. For example, the exclusion of topics like race and eugenics and the pragmatic balancing of
human research guided by the ethos of animal research were driven by
a combination of judgments about scientific tractability and political expectations. Yet the general orientation of this combination was to define
capital so as to head off controversy and thus to secure behavior geneticists' autonomy to guide the field.

"We Considered Omitting Human Studies Completely"

The long shadow of eugenics meant that behavior geneticists had to reconstruct the scientific archive. As Fuller and Thompson wrote in the
preface to their seminal 1960 textbook, "Another difficulty has been in
selecting suitable materials for citation, especially when these fell short
of exacting scientific standards. . . . We considered omitting human studies completely, but did not do so because of the primary interest of many
readers in human problems."[69] Not surprisingly, just as animal researchers were the key field organizers, their research predominated during the
field's early period (see table 2.1).

Behavior geneticists understood that much of the interest in their science would turn on its capacity to illuminate human behavior, so it would
not do, at least in the long run, to justify an interest in animal behavior
"for its own sake." As Scott explained, "From the very outset of my interest in animal behavior, I saw that if I and other researchers expected
substantial financial support it would have to be associated with some

TABLE 2.1. **Animal and human research cited in review articles**

Reviews	Animal research articles	Human research articles	Articles on other topics
Hall (1951)	43	1	8
Fuller (1960)	84	29	17
McClearn and Meredith (1966)	134	93	18
McClearn (1970)	92	93	17

Note: These reviews have almost no overlap in the articles they consider. "Other topics" include methods, general topic works (e.g., textbooks about genetics), or subjects that could not be determined from titles.

major practical problem or problems."[70] But the motive to bring animal and human research together was intellectually justified as well. Fuller and Thompson said that beyond readers' interest, "the juxtaposition of animal and human studies also has value, we believe, in the search for general laws of behavior."[71] Psychiatrist Eugene Bliss wrote that one might reasonably "reject the entire field [of human behavioral research] as hopelessly befuddled." So the research committee of the American Psychiatric Association sponsored "a meeting devoted to animal behavior, not because the ultimate answers to man's behavior were evident in contemporary work in this field, but because the area offered new techniques to study problems of behavior."[72]

The reason for the problematic status of human research was its focus on "applied problems." In introducing their chapter on human studies, Fuller and Thompson offer this apologia, "The desire to put the newly discovered science of genetics at the service of human welfare led some early twentieth-century scientists to make excessive claims for the importance of heredity in the origin of social maladjustment."[73] Thus for many behavior geneticists the key to unifying the field and searching for "general laws" was less about finding animal analogues for relevant human behaviors than bringing the "pure research" ethos of animal scientists to human research.

The focus of a "pure science" study of human behavior, as Jerry Hirsch articulated it, had the "objective of the discovery and understanding of natural units."[74] Research should be focused on "the simplest possible units of intellectual functioning among individuals of known ancestry."[75] Examples of these simple units included autonomic nervous function, memory span, and sensory discrimination of taste, color, sound, and touch. He hoped the identification of behaviors that were (arguably) rel-

atively close to biological control combined with family data would enable their genetic dissection. In contrast, "omnibus tests of personality
and behavior . . . may be useful instruments of classification to serve the
practical needs of society . . . [but] we should not expect them to be very
precise measures of biological differences."[76]

While such views were common among animal behavior geneticists,
there was some dispute about the strategy among human researchers. At
a discussion during the 1965 Louisville conference on human behavior
genetics, Robert C. Nichols remarked: "The criterion of social importance is one that we should look at in human behavior genetics; the only
reason I can see for working with human behavior at all, is its importance to people. . . . Trying to find simple measures because these may be
closer to the genetic mechanism doesn't seem to be a compelling argument. If this is your interest, you should not be working with behavior in
the first place."[77] But other human researchers at the same event agreed
with the position Hirsch's view exemplified. Ruth Guttman argued that
studying "trivial traits" like eye color and bristle number in flies had historically led to genetics' intellectual leaps.[78] Students of human psychology should not worry that trivial traits will mean their research has trivial implications. Medical geneticist Gabriel Lasker suggested focusing
on known simple genetic changes affecting physiology (like PKU) and
then giving a large battery of psychological exams.[79] Raymond Cattell's
solution to the somewhat arbitrary character of psychological tests was
not physiological reduction but mathematical abstraction—using statistical factor analysis to identify common structures unifying psychological tests.[80]

Before 1970 most behavior geneticists thus imagined a symbiotic relationship between animal and human research. Concern with human
problems justified interest in the topic, while the pure science ethos of
animal research would firm up the field's scientific status and lead to
general laws of behavior. Some human researchers bristled slightly at the
suggestion that they needed to adapt the "pure science" ethos, but a tolerant attitude reigned. Field members were generally interested in opening up new approaches, and they weren't closing down ones they considered less fruitful (apart from race differences, eugenics, and crude
determinism). This symbiotic relationship also entailed an integration
and a low division of labor between animal and human researchers.
Since each had implications for the other, behavior geneticists evaluated,
interpreted, and sometimes researched both.

Attacking Behaviorism, Refining Individualism

An important collective motivation of behavior geneticists was to get the social and behavioral sciences to take genetic causes of behavior seriously. The inclusion of additional genetic variables was insufficient; rather, their aim was overthrowing the dominant paradigm of behaviorism. Behavior genetics owed a good deal to behaviorism. For one, it shared its empiricism and rejection of introspective methods and its skepticism about mind or the unconscious as meaningful psychological entities. Further, behavior geneticists took from it the basic concept of "behavior," or the concern with organisms' actions and abilities rather than their "character" or "quality," which had been the focus of eugenics. But they sought to correct the neglect of inheritance in behaviorism through a detailed critique of the notion that behavior can be explained solely in terms of conditioned responses to environment.[81]

The objection was sometimes framed in terms of the egalitarian view of humanity in behaviorism and environmentalism more broadly. Behavior geneticists were (and continue to be) fond of scoffing at the statement by John B. Watson, father of behaviorism, that he could take a dozen babies and train them at random to be any profession from doctor to thief.[82] A good example can be seen in psychologist Sandra Scarr's account of her initial interest in behavior genetics that formed in reaction to "the social scientific view of the time [which was] . . . that all individuals and groups were equally endowed with everything important, such as genes for intelligence (whatever those might be). My own observation about human differences made me curious about the . . . certainty on this matter, particularly when I noticed the lack of evidence for such a view."[83] Not all were so flip: though Jerry Hirsch called behaviorism a "fiasco," he also said that Watson knew he was "going beyond the facts" in claiming babies' unlimited malleability because he was reacting to hereditarians' equally extreme claims about determinism and racially fixed behavior.[84]

The glib rejection of egalitarianism (with politically conservative overtones) is often how behavior geneticists' critique of behaviorism is remembered. But there was a more important, scientifically motivated concern that was to direct behavioral scientists to take variability more seriously. As behavior geneticists like Jerry Hirsch argued, one pernicious consequence of assuming no individual differences is that when behaviorists observed variation under given environmental conditions they implicitly assumed that it was error around the "true" average value.[85] Once the average is obtained, it is easily reified as characteristic of the group (and the individuals within it). Then it is an obvious step to compare the groups (strains of lab animal, genders, classes, races, and such) as if they are individuals.[86] Hirsch explained this chain of logic with the example of Henry E. Garrett, a hard hereditarian and former president of the American Psychological Association, who in the 1950s and 1960s helped organize a campaign against racial integration. Attacking an argument Garrett made in *Science*, Hirsch wrote, "He ignores individual differences and claims that wherever two populations differ on some scale of measurement, no matter how vague, any individual from the population with *the* higher mean is better than any individual in the other population, and that intermarriage will 'be not only dysgenic but socially disastrous'!"[87] Although Garrett himself railed against the "equalitarian dogma" of social science, Hirsch showed that his racist claims were symptomatic of the typological style of thinking and misunderstanding of variability characteristic of behaviorist psychology.[88]

For Hirsch and others, the basic problem with behaviorism was the quest for "laws of environmental influence." For "law" to have any meaning here, researchers would have to assume that effects observed were universally true. For them to be universally true, they had to assume that all individuals reacted the same and, once again, that differences or variations were noise or error. But as Hirsch forcefully argued, meiosis makes individuals radically unique genetically.[89] Further, in all cases where proper genetic and environmental controls had been made, phenotypes were always the product of unpredictable interactions: "it is ridiculous to attempt to characterize an environment as generally favorable or unfavorable, or any genotype as generally superior or inferior. Some average measure of an environmental influence is applicable *only* to those genotypes affected by it in the same way. Similarly, any rank ordering of genotypes can be applied only in those environments which preserve the ranks of the phenotypes."[90] The "error" that behaviorists observed was actually genotypes responding differently to particular environmental conditions. "Since genotypic diversity and genotype-environment interaction are apparently ubiquitous," Hirsch wrote, "attempts to study the laws of environmental influence have been grasping at shadows."[91] But, it should be noted, in this Hirsch was equally dismissive of genetic determinist or racist interpretations because "the characterization of a genotype-environment interaction can only be *ad hoc*."[92] The performance of a genotype in one rearing or testing situation can never be generalized into another unobserved one; a claim that has clear implications for education.

An important, though subtle, aim of behavior geneticists during this period was therefore to focus attention on a few issues regarding individuality and groupness. First, groups should be conceived as populations that are aggregates of individuals unified by some objectively specifiable genetic or environmental conditions. Groups (or behaviors) should not be defined a priori by society—"there is no reason to think society knows what a variable is," as Gardner Lindzey put it.[93] Further, the key property of populations is variation within them, and the point of genetics is to explain the genetic and environmental causes of that variation. Thus focus should be maintained not on the properties of the population so much as the differences among the individuals that comprise it. But the focus should not go all the way to the individual per se. Behavior geneticists did not see themselves as having traction on the question of what caused an individual behavioral outcome, especially not for humans. Hirsch was particularly occupied with being precise on this point: "we have come to realize that it is impossible to study the genetics of behavior. We can study the behavior of an organism, the genetics of a population, and individual differences in the expression of some behavior by the members of that population."[94]

Beyond Nature versus Nurture

There was a crucial risk, however, to identifying individual differences as a key object and analyzing genetic and environmental sources of variation as a core task. The danger was that the field would unwittingly revive and become entangled in the "heredity/environment" controversy, which they already considered themselves to have "won." Even if they hadn't convinced all environmentalists to adopt genetic methods, behavior geneticists had demonstrated the reality of individual differences and the importance of heredity on behavior. The obvious next step was to measure which factor mattered more to behaviors.

Already in the 1950s proto-behavior-geneticists were eager to push past this issue. In his 1951 review, Calvin S. Hall looked forward to the day when psychogenetics would "free itself from distracting excursions into pseudo-problems, chief among them the heredity-environment issue."[95] In her oft-cited 1957 presidential address to the American Psychological Association, Anne Anastasi said that "the traditional questions about heredity and environment may be intrinsically unanswerable. Psychologists began by asking *which* type of factor, heredity or environmental, is responsible for individual differences in a given trait. Later, they tried to discover *how much* of the variance was attributable to heredity and to environment."[96] She claims that these efforts must be put aside, and researchers should ask *how* genes and environments produce behavioral differences.

[margin handwriting: Not which or how much... but how]

Not all behavior geneticists agreed that vaulting over the "how much?" question was desirable. In his 1960 review, John Fuller disagreed that the question was obsolete: "In dealing with populations . . . the contribution of heredity to total variance is still a useful object of inquiry."[97] A key reason for Anastasi's rejection was that she considered "traditional investigations . . . to yield inconclusive answers."[98] But there were new datasets and methods that behavior geneticists were eager to apply to the problem.[99] These researchers were confident that with much better measures of people's behavior; much larger and more rigorously collected data about twins, adoptees, and families; and emerging statistical models (coupled with increasingly powerful mechanical computers) that much higher quality estimates of heritability were now possible.

But many behavior geneticists were wary of staking too much on heritability estimates. Heritability is the proportion of variance for a trait due to genetic variance. As many pointed out, however, "heritability is a property of populations and not of traits."[100] It is always linked to a specific group of individuals, with a particular distribution of genotypes experiencing a particular range of environments. The measure is always local in this way, so talking about the heritability of intelligence or personality disconnected from these particularities is misleading. Heritability is local in another way: "one can *not* infer from a high heritability that environmental selection is hopeless."[101] Within a different set of environments, the heritability of the trait might be different, but the phenotype itself might change in other ways. Intelligence in a population raised under different conditions might be higher overall or distributed differently. That heritability had no bearing on such issues had been demonstrated both experimentally (with rodents) and theoretically.[102] For these reasons Hirsch declared that heritability estimation could be an initial research step, but "I cannot seriously consider heritability to be 'one of the central concepts of modern genetics.'"[103]

[margin handwriting: Don't want to look @ populations]

At a less analytic level, some behavior geneticists worried that emphasizing heritability estimation would undermine the field's seriousness. In his textbook focusing on animal research, P. A. Parsons complained, "unfortunately most of the work on behavioural traits so far, has been aimed at showing them to be under genetic control without further elaborations."[104] Human psychologist Daniel Freedman wrote that "it has been my impression that behavior geneticists have a gimmick rather than a theory."[105] He argued that they were successful at criticizing environmentalists but had not replaced their ideas with something more substantial. This anxiety threatened to stain the young field's reputation more broadly. The popular science magazine *Scientific Research* covered a 1966 conference on the genetics of human intelligence at Rockefeller University. The article labeled psychologist Steven Vandenberg's work a "debacle" unable to define intelligence or distinguish environmental from hereditary causes. NIH psychologist David Rosenthal's mocking criticism was covered in detail: Rosenthal "composed a litany of some of their blatantly humorous results. 'High heritability—' he began the intonation, 'sang in a glee club; low heritability—took voice lessons. High heritability—rode roller coaster; low heritability—rode sports car.' Here the audience laughed."[106]

Transdisciplinary Breadth

Many behavior geneticists were thus anxious that the field, especially human research, would become overly focused on generating heritability estimates. A range of research agendas alternative to or complementary to heritability estimation had been proposed. Their aspiration was not to impose a particular paradigm or to draw boundaries around a particular set of disciplinary practices but to create hybrid intellectual space where a range of questions, data, and disciplinary approaches could be pursued. Some examples will provide the flavor of these proposals and highlight their cross-disciplinary breadth and ambition.

One line involved the proposal discussed above that human behavior genetics, in a "pure" science mode, should study behaviors closely related to physiology. Generally this was related to an idea that behavior genetics should more closely align with medical genetics and population genetics. For example, behavioral effects might be studied in people with conditions or diseases due to specific genetic defects.[107] Behavior

might also be studied in genetically isolated communities (the Menno-
nites, Old Order Amish, and Otomi Indians of Huixquilucan, Mexico,
were proposed) where traits could be traced through pedigrees.[108]

Another range of proposals concerned the exploration of how heredity and environment interact to produce behavioral differences. Among
Anastasi's proposals included animal breed comparisons, studies of prenatal and early rearing experience, and the specific developmental context of twins.[109] Yet another set concerned genetically caused physical
traits or disabilities (retardation, deafness, ill health, or physique, for
example) and the range of direct and indirect effects on the individual's behavior. One example concerned how physique affects psychology
through one's sense of effectiveness as well as treatment by others.[110]
These ideas put behavior genetics at the nexus of biology, social interaction, and culture. Psychologist David Rosenthal gave an example of
how to study social introversion by comparing new families with different arrangements of parental introversion, then carefully observing parental behaviors and babies' development over time as well as responses
to experimentally introduced social and nonsocial stimuli.[111] He also advocated more attention to environment rather than simply controlling it
with twin and adoption designs. This meant both carefully measuring
particular aspects as well as comparing radically different environments,
such as nuclear families and collective rearing in kibbutz settings.

Still other proposals focused on large-scale cross-cultural comparisons. For example, one of Anastasi's proposals was for comparisons of
rearing practices in different cultures. Animal behavior geneticist Benson Ginsburg and anthropologist William Laughlin laid out a massive,
ambitious agenda for exploring "the multiple bases of human adaptability and achievement." They said, "comparative studies must attempt
to sample the total range of human capacities and should include both
similar and dissimilar genotypes in as widely varying environments as
possible."[112] By exploiting the amazing global ubiquity of humans and
the multiplicity of physical environments and social orders to which they
have adapted, they believed scientists would truly be able to understand
the origins and possibilities of human behavior. Others advocated focusing on universal human behavioral traits, for example, linguistic ability, or analyzing behaviors in a framework of evolutionary adaptation.[113]
And another group of behavior geneticists were eager to analyze not individual traits but social behaviors and the behavioral reactions of subjects (usually animals) to particular situations.[114] Such studies would in-

tegrate behavior genetics with research in anthropology, population genetics, ethology, linguistics, and evolutionary theory.

Behavior geneticists during this period were thus very interested in promoting an ambitious and broad intellectual agenda. Indeed, their harshest words for each other were about avoiding narrowing of the agenda. Thus the warnings about the dangers of heritability estimation were animated by the fear that it would be easy to become fixated on it. And the defenses of heritability estimation were made on the same grounds of preserving a broad tool kit and not foreclosing scientific possibilities. The most acute anxieties expressed were about the perception of a collective failure to explore different approaches broadly and quickly. As Rosenthal lamented, "I find the amount of progress disappointing . . . suggestions for further research have not been pursued."[115] This is the kind of complaint restlessness and a sense of boundless possibilities produce.

Conclusion: An Implicit Pact

These developments show that the field of behavior genetics was founded in the shadow of a controversy that didn't erupt into active dispute: How would it be possible to study the heredity of behavior scientifically given its explosive historical legacy and political potential? Compared to the primary tasks of field building—assembling a network of interested researchers, training new members, and building the archive of methods, concepts, and findings—avoiding this controversy was a secondary concern. Yet as we have seen, avoiding controversy strongly shaped the ways behavior geneticists pursued their field-building tasks.

Most crucially, the field's leading architects were animal behavior geneticists whose research was removed from applied topics that were most amenable to politicization. They generally agreed that the ultimate point of the field was to address human problems, and that this was necessary to sustain interest (and funding). However, they saw the path to this goal starting with a "pure science" ethos committed to objective definition of traits, analytic precision, and focusing on the mechanisms that produce behavioral differences. Along these lines, behavior geneticists pitched their efforts at solving academic problems and changing the attitudes of academic researchers. They weren't seriously drawing the attention of media sources or entering into dialogue with a policy agenda.

Behavior geneticists worked hard to assemble a network of scholars, especially in the 1960s, with conferences, workshops, and training sessions. They enthusiastically included scholars in a range of disciplinary backgrounds—genetics and psychology most prominently, but also anthropology, sociology, psychiatry, demography, zoology, agricultural sciences, and statistics. But they were also cautious about their associations; they kept their distance from hard hereditarians, racial claims makers, and traditional eugenicists. At the same time they worked closely with geneticist Theodosius Dobzhansky, whose positions against scientific racism and eugenics were well known. Behavior geneticists also worked to diffuse the potentially controversial character of the field in the topics they studied. Part of this involved pointing out the scientific shortcomings of hard hereditarianism, eugenics, and racial research and clearly differentiating their work. It also meant articulating a broad range of research questions and methodologies that integrated many disciplinary perspectives.

We can see in these actions an implicit pact among behavior geneticists. As a condition of participation they agreed not to promote genetic determinism, engage in racial comparisons, or pursue social policy implications that might be perceived as eugenic. Indeed, almost never did researchers even raise fatalistic interpretations of their work—that genetic causes of intelligence might limit educational possibilities.[116] The agreement to avoid these issues was partly political, though somewhat indirectly. This was not about particular political commitments so much as the desire to avoid politicization and controversy. But more crucially, avoiding these issues was part of a scientific calculation. Behavior geneticists did not believe racial comparisons or eugenics were inherently illegitimate. Rather, they could only be considered under certain conditions that were impossible to meet in the present. Eugenic policies could only be effective in a society that already maximized people's capacities. The technical problems of making racial comparisons swamped their scientific or social interest.

There was remarkable consensus on these points among behavior geneticists. Although individuals calling for these forbidden topics certainly existed (we will turn to some of them in the next chapter), they were not considered part of the field. To the extent that behavior geneticists harbored sympathy for such views, they kept it quiet.

The points of dispute in behavior genetics were at a completely different level. How close should human research be modeled on the animal paradigm? Should researchers focus on behaviors close to physiology or aggregate traits like personality or intelligence? How important would generating heritability estimates be? Which new research directions should be pursued? To be a behavior geneticist meant being committed to these kinds of questions. These were questions that were given by the practical intellectual problems a community of researchers faced. The issues debated were, at heart, academic. If the field's members aspired one day to aid in solving social or medical problems, they believed that this would only happen by getting the science right first and being precise about what it means.

The pact among behavior geneticists had other dimensions as well. Animal and human researchers relied upon each other for legitimation. Human researchers needed animal researchers to provide scientific legitimacy and to help rehabilitate the topic from its disreputable past. Animal researchers needed human researchers to sustain interest and funding in what they were doing. To credibly build a universal science dedicated to uncovering the laws of behavioral inheritance, animal and human researchers would have to work together as would scientists from a wide range of disciplines.

The pact also involved mutually supportive engagement. This wasn't just a field of mutual interest where scientists from different disciplines or with different approaches worked in parallel. Behavior geneticists were trying to build an integrated science—each of eight major edited collections from this period exhibits substantial disciplinary and methodological diversity.[117] The tone of these efforts is mutual encouragement, criticisms are usually mild (and generally involve exhortations to increase the creativity and variety of research), and tensions between disciplines are muted.

There was also a sense of unity produced by a common enemy. The figure of the behaviorist or environmentalist served as a kind of unifying totem. Indeed, its contradictory character confirms its symbolic, totemic status. After all, in the same breath that radical environmentalists were being denounced as opponents, behavior geneticists would say the influence of heredity on behavior was beyond dispute by responsible scientists. In other words, their enemy was a figure they claimed didn't actually exist.

The period prior to 1970 was a kind of scientific golden era for behavior genetics. Behavior geneticists appeared to have successfully headed off destructive controversies and put the field on solid scientific ground. Scientists from many backgrounds and with many interests were flocking to the field to participate. Geneticists and psychologists predominated, but researchers from across the social, biological, and medical sciences eagerly participated too. No single methodological or theoretical approach dominated, and researchers eagerly proposed creative ways to study genetic influences on behavior and the interactions between heredity and environment.

Behavior genetics was closest to becoming the universal field of the "genetic inheritance of behavior," which I discussed in the introduction, during this era of its initial formation. Behavior genetics was relatively integrated. The different disciplines, questions, methods, and research subjects (animals and humans) were thought to have important implications for each other and high value in their juxtaposition. Some behavior geneticists even began to make the argument that heredity need not be a particular focus of the field. Environments could be studied more fully and integrated with genetic information to produce a comprehensive and balanced picture of the causes of behavior.[118] Or the focus could be on inheritance more broadly and the set of genetic, cultural, and material elements passed from generation to generation that affect behavior.[119] These differences indicate that behavior genetics had no consensus or paradigm, but rather the conditions were in place for a healthy and diverse competition for scientific capital.

But the future as an integrated transdiscipline was not to develop for behavior genetics. The controversy about racial differences in IQ was about to explode onto the academic scene, completely enveloping behavior genetics and deflecting its course of development. Its membership and scientific agenda would never again be so balanced and integrated. Its atmosphere of openness and cooperation would be replaced by hostility, paranoia, and a bunker mentality. Its scientific bona fides would be challenged, and politicized science would become a permanent feature. After Arthur Jensen's intervention in 1969, behavior genetics' golden era was over.

CHAPTER THREE

The Young Field Disrupted

The Race and IQ Controversy

The late 1960s were filled with great hope for behavior geneticists. They had assembled a small but enthusiastic membership, were in the final stages of setting up the field's society and journal, and had articulated an open research agenda they believed would move them beyond the limits of the nature versus nurture debate. These and other efforts had the effect of forming an implicit pact that would bind together the interdisciplinary field while distinguishing it from eugenics and race research, which had undermined previous efforts to link genetics and behavior.

But this settlement was disrupted before behavior geneticists were able to build on these accomplishments. In 1969 the Berkeley psychologist Arthur Jensen drew heavily on behavior genetics in his infamous article "How Much Can We Boost IQ and Scholastic Achievement?"[1] It argued, first, that the IQ gap between the black and white populations has genetic causes and, further, that the genetic roots of IQ mean that educators can do little to improve it or reduce inequality. His use of behavior genetics drew the field into the crossfire of disputants in the IQ controversy. By the late 1970s, when the IQ controversy had cooled, Jensen and his ideas had been incorporated into behavior genetics. This meant the reversal of some of the most crucial terms of the implicit pact that had previously organized the field. Where previously they had denounced racial research as scientifically intractable and socially destructive, by the end of the controversy they agreed that race research was socially important and that genes do influence the IQ gap.

There are two obvious ways to explain this reversal. The first is that behavior geneticists actually believed in the genetic inferiority of blacks, for basically political reasons, and that the controversy gave them the opportunity to generate ad hoc scientific rationalizations. The second is that behavior geneticists had a set of scientific commitments that entailed incorporating Jensen's arguments. Thus the disputes between behavior geneticists and their critics were driven either by political differences or different conceptions of science.[2]

My explanation is different: The incorporation of Jensen was a product of behavior geneticists' efforts to defend their field and preserve their capacities to define it with as much room for maneuver as possible. As I will show, a small number of behavior geneticists saw Jensen's intervention as evidence that the terms holding together behavior genetics were unsustainable. Their solution was vigorously rejecting race research and many of the concepts Jensen had deployed—which would have effectively eliminated human behavior genetics. Most behavior geneticists, however, aimed to stick together and position the field in a middle ground between Jensen and his critics. But as the controversy unfolded, critics increasingly attacked Jensen and behavior genetics together. As critics boxed them in with Jensen, behavior geneticists developed scientific and political arguments as well as views of their critics to push against the contraction of the field. In the course of these efforts, Jensen and his views were incorporated while the critics were bounded out of the field. Analytically, the field approach insists on the idea that the actors' "interests"—what counts, often implicitly, as sensible, reasonable, or rational behavior—are mediated by the evolving struggles and forces comprising the field. This chapter shows in detail how the IQ and race controversy changed the field's logic of practice. Thus while behavior geneticists and other actors pursued their interests during this controversy, these interests were being redefined, fractured, and reprioritized.

The story of how Jensen's race science became incorporated into behavior genetics occupies the middle of the chapter. Before it, I consider why the IQ controversy appeared from the start as a massive academic and public scandal that was so difficult for behavior geneticists to handle. Jensen followed a string of efforts by other scientists—most notably William Shockley, the Nobel physicist turned eugenicist—who had drawn great attention, but little scientific legitimacy, to the idea that science could prove blacks' inferiority. Jensen's contribution was so explosive because it suddenly seemed to offer scientific credibility to these provocative ideas.

The final sections detail the effects of behavior geneticists' handling of the IQ controversy beyond the incorporation of Jensen. Behavior geneticists and critics became incredibly hostile toward each other. While behavior geneticists were able to close ranks against their critics and dampen their impact within the field, this had unintended consequences. One was that behavior genetics' ties to the discipline of genetics loosened as critics' polemics discouraged geneticists from engaging the field. Another was that the scientific agenda of behavior genetics became entangled in a nature versus nurture framing and the defense of the genetic explanation for racial differences that was to reverse the crucial terms upon which the field's founders had hoped to establish it.

Consequences: ① lose connection w/ genetics ② Entanglement in nature vs nurture

Attack on Racial Egalitarianism in Science

The story of the unfavorable terms under which behavior geneticists were compelled to take up the matter of race differences in the 1970s begins with the postwar scientific orthodoxy on race. Scientists were horrified by the role of scientific claims in the execution of the Nazis' racism and mass murder. In reaction to that legacy, scientists from many fields articulated a new orthodox view on race. Among the best known of these efforts were the UNESCO-sponsored Statements on Race in 1950 and 1951 that denounced the idea that science proved any kind of racial hierarchy or that racial intermixing represented any kind of danger.[3] While there was some disagreement about the causes of racial inequality between social scientists, who emphasized environmental differences, and geneticists, who were generally agnostic because they viewed the question as methodologically intractable, racial egalitarianism animated the Statements on Race and the views public intellectuals propagated more broadly.[4]

Major claim made by science

Burned by eugenics, geneticists nurtured their scientific authority after the war by retrenching from the politics of race.[5] Social and behavioral scientists were doing the opposite. They were building their authority by working to strengthen American democracy, and they saw racial integration as a major component. Thus dozens of leading social scientists and psychologists testified to the harm of segregation on black children in *Brown v. Board of Education* as expert witnesses or via amicus brief.[6] Further, they became intellectually and professionally invested in the justification, design, and implementation of desegregation efforts and Great Society social programs.[7]

Most moved away from the politics of race

Tension: some sciences moving away while others moving towards in order to support better living

The rising political stakes of this scientific research created demand for scientists who would challenge racial egalitarianism. A small cadre of psychologists and biologists eagerly filled this niche and attacked what they called the "equalitarian dogma." They argued that desegregation was based on the premise of blacks' equal potential, which they disputed. Further, they claimed desegregation would produce miscegenation, which was biologically dangerous.[8] But this group had trouble drawing attention. Other scientists largely ignored or dismissed them as biased and unscientific. And apart from the occasional *U.S. News and World Report* article and plaudits from hardcore segregationists they got little traction with the public. But the fortunes of antiegalitarianism would rise with the advent of William Shockley.

Shockley had won the Nobel Prize in Physics in 1956 for coinventing the transistor, but in the mid-1960s he turned his attention to eugenics and race. He delivered a series of public lectures and interviews on "reducing the heredity-environment uncertainty" regarding racial differences in IQ and crime and possible eugenic solutions to "human quality problems."[9] Among his proposals were requesting that black public figures submit blood for genetic tests to see if their eminence could be explained by their proportion of "white genes" and the "voluntary sterilization bonus plan," which was a "thought experiment" about the effects of paying low IQ individuals to elect sterilization.[10] He also mounted an extended campaign to get the National Academy of Sciences (NAS) to "ignore racist implications and institute studies on the effects of genetic inheritance upon human behavior relative to city-slum problems."[11] The result, a report issued in 1968 by three eminent geneticists, James F. Crow, James V. Neel, and Curt Stern, concluded:

> In the absence of some now-unforeseen way of equalizing all aspects of the environment, answers to this question [of hereditary differences in intellectual and emotional traits between races] can hardly be more than reasonable guesses. Such guesses can easily be biased, consciously or unconsciously, by political and social views.[12]

If Shockley had moved the egalitarian position, it was in the "wrong" direction. Where the UNESCO statement of 1952 said there were no good data on racial differences, the NAS statement said that such data would be essentially impossible to get without the achievement of social and political equity.

Thus in the short run Shockley failed in his goal to raise the scientific credibility of antiegalitarianism. Yet his impact was profound. First, he got the question to be considered by elite scientists. Second, while his predecessors had argued the antiegalitarian position in terms of physiology, evolution, education tests, and so forth, Shockley centered the racial differences problem on behavior genetics, building his case with its methods and data and arguing for greater research in the area. Further, with dogged determination he kept exploiting the inherent uncertainties in the area to push outcomes he preferred. Denouncing the 1968 report as "Lysenko-like," Shockley kept pushing the NAS until in 1972, well into the Jensen controversy, a second committee tepidly agreed that "investigation of the nature and significance of individual, populational, and racial hereditary differences in the human species is a proper and socially relevant scientific subject."[13] In addition to forcing elite scientists to pay attention, Shockley also drew broad public interest. His Nobel status combined with his frequently outrageous remarks made him irresistible to journalists. His talks often sparked protests and picketing by enraged radicals, but Shockley understood this attracted attention and often a degree of sympathy to him.

During this campaign Shockley sacrificed most of his scientific credibility. His outrageous proposals and the dogged way he pursued them led his peers to perceive him as a crank—a once brilliant man whose obsession with racial differences led him to sacrifice his scientific judgment. Perhaps for this reason behavior geneticists never felt it necessary to address his claims in more than passing mention despite the fact that he constantly invoked their work in his campaign.[14] If advocates of antiegalitarianism had continued in the path of Shockley and his predecessors, the history of the topic and of behavior genetics might have been very different. But partly due to Shockley's encouragement, Arthur Jensen would enter the scene bringing new life to the topic.

Jensen was a Berkeley educational psychologist who, unlike those preceding him, had special expertise in questions of intelligence, learning, and racial differences. He had worked entirely in an environmentalist mode, focusing on black children's "cultural disadvantages" in education.[15] Bothered by anomalies he felt environmentalism could not explain, Jensen turned to hereditarianism. His watershed contribution was the 1969 *Harvard Educational Review* article "How Much Can We Boost IQ and Scholastic Achievement?"

In a soberly and methodically argued 123-page review and synthe-

Arguments

IQ = fair

IQ = 80% herid.

IQ blacks 15 pts ↓ based on genes

sis of literature from the fields of intelligence testing, educational psychology, sociology, population genetics, and behavior genetics, Jensen claimed that compensatory educational programs—like Project Head Start, which aimed to improve the school performance of poor and minority children with intensive public preschool—were destined to fail. His basic argument had several parts. First, he argued IQ scores are a valid and fair measure of intelligence that accurately predict educational and social success. He then claimed that genetic studies show that IQ is about 80 percent heritable—this means that if we look at the variance in children's IQ scores (not the average score of the population but their distribution), 80 percent of the differences in those scores are due to differences in the children's genes, and only 20 percent of the differences are due to differences in the environments they experience. Genes, he contended, are much more important than educational differences in determining which children are more intelligent and thus more socially successful.

Further, the average IQ of blacks was about fifteen points lower than that of whites, and Jensen argued that efforts to explain this gap environmentally had failed. For example, when factors like income, school quality, and family structure are controlled statistically the gap narrows but remains. And children of relatively high-status black families had lower IQs than disadvantaged white children, while disadvantaged black children did worse than American Indian children who had even worse measures of disadvantage. When the high heritability of IQ in whites, the genetic heterogeneity between the black and white populations, and the inadequacy of environmental explanations of the IQ gap are combined, Jensen wrote, "All we are left with are various lines of evidence, no one of which is definitive alone, but which, viewed all together, make it a not unreasonable hypothesis that genetic factors are strongly implicated in the average Negro-white difference."[61]

Shockley and others had drawn tremendous attention to the genetic hypothesis for racial differences, but their tactics tended to confirm the view that their ideas were just warmed-up racist eugenics. Jensen's article transformed the debate by capitalizing on the public interest Shockley and others stirred while presenting the genetic hypothesis for racial intellectual differences in a way that made it appear eminently *scientific.* This was largely a difference of stylistic mastery grounded in Jensen's expertise as an educational psychologist. The article, after all, was a review that presented only materials that were already available. However, where Shockley and others had presented the case partially, suggestively, provocatively, and politically, Jensen made it in a detailed, balanced, dispassionate manner that seemed unmotivated by the politics of segregation or racist anxieties. Further, his predecessors' interest in racial differences had been largely avocational, which called their motives into question. Jensen's interest was embedded in a psychological and educational paradigm, and his conversion to hereditarian theory aimed to solve certain scientific problems he believed unanswerable from the cultural disadvantage paradigm under which he had previously worked. This trajectory gave Jensen special credibility, but it also meant that he was like a fish in water, scientifically speaking. He made the strongest possible case by linking his argument to the many relevant scientific literatures, and as the debate raged over the next few years, Jensen was prepared to respond to all challenges.

The media attention to the controversy was intense and enduring. The IQ controversy became perhaps the most widely covered academic controversy ever—dozens of articles and television programs appeared in major and minor media outlets.[17] Parts of the debate were conducted through the popular media in addition to traditional academic venues, and each side of the debate accused the other of misinforming the public.[18] Radical student and civil rights groups frequently protested what they saw as the racist and classist implications of the research. At many campuses in the United States and United Kingdom they picketed, protested, and disrupted lectures and teaching by Jensen and his supporters, occasionally with violent results.[19] The overarching impact of the media attention and the protestors was to produce an aura of scandal around the IQ debate and to highlight its high stakes, which undermined the ability of participants to approach the debate in purely academic terms. Substantively, the debate focused heavily on the clarification of terms, and this was partly due to the fact that the technical and commonsense understandings of terms like *heritability* and *genetic determination* diverged, yet public attention kept both sets of meanings circulating. Finally, this public attention sowed distrust among debating scientists, since each side thought the other was misleading the public, and Jensen and his supporters considered the critics partly culpable for the treatment they got from protesters.

The academics who responded to Jensen can be divided roughly into four categories. First was a small group of his allies who weighed in to the controversy supporting most or all of what he said, sometimes ad-

Groups

① Allies

② people who had to respond b/c of their field

③ Critics w/ issues & central to BG

④ BG's

vancing or elaborating it. Among the most noteworthy were Shockley; Hans Eysenck, a British psychologist whose *IQ Argument* popularized Jensen's argument and emphasized the race claims; Richard Herrnstein, an animal psychologist who converted to behavior genetics research— initially agnostic about race, he argued that genetic differences explain class structure; and Raymond Cattell and P. E. Vernon, both eminent hereditarian psychometricians.[20]

The second category was comprised of the large and diverse set of scholars provoked to respond because Jensen had drawn on their fields or expertise to make his case. Thus many psychologists, sociologists, anthropologists, educationists, statisticians, policy researchers, and biologists engaged in debates about the reality and measurement of the IQ concept, how IQ tests might be biased and how they should be used to make educational decisions, whether compensatory education had truly failed, the history of discrimination, whether race is real, and so forth.[21] Although much of the IQ debate circulated around their concerns, for the story of behavior genetics, which is my focus, their main function was to add to the intensity and complexity of the academic scandal and to raise the stakes for behavior geneticists. Jensen had placed behavior genetics at the nexus of their concerns and thus positioned it to challenge multiple lines of scientific authority. He established this broad group as an audience standing in implicit judgment of behavior genetics, which contributed to the energy of the controversy.

The third group was Jensen and his allies' dedicated critics who focused on issues central to behavior genetics. Richard Lewontin, a population geneticist, and Leon Kamin, an animal psychologist, are among the best known, and I return repeatedly to their interventions. However there were many others, including neuroscientist Steven Rose, geneticist Jonathan Beckwith, evolutionary theorist Stephen Jay Gould, sociologist Christopher Jencks, physicist David Layzer, economist Arthur Goldberger, and members of radical science groups like the Boston Sociobiology Study Group and Science for the People.[22]

The fourth group, finally, was the behavior geneticists. And we turn now to their difficulties responding to the controversy in the context of these other parties.

The Incorporation of Jensen *Didn't know how to handle it*

The controversy surrounding Jensen was an unprecedented problem for behavior geneticists and their young field. On the one hand, Jensen had used a variety of behavior genetics' tools, data, and findings; presented them in a forceful, creative, and unsettling way; demonstrated their relevance to multiple audiences; and given them the kind of wide academic and public exposure that few scientists could even dream about. But on the other hand, he had aligned behavior genetics with the race differences topic, offered a basically determinist interpretation, invoked eugenics, and had made bold public policy recommendations that politicized the science. Not only did Jensen directly challenge the implicit pact, discussed last chapter, that behavior geneticists hoped would ground their field's success, but he also did so in the most public, scandalous way possible. And thus behavior geneticists were rightly worried about being attacked—both intellectually and physically—about losing their funding, about the coherence of the research community, and about "supporting a field that's inhuman in the long run," as one interviewee put it.[23]

Today, the association between behavior genetics and Jensen and his allies has become such an article of common sense that, apart from the few field members who have been critical, neither behavior geneticists nor observers of the field have seriously questioned it. There seems to be a presumption that intellectual and political affinities made the link fait accompli. I will show, however, that the incorporation of Jensen was a contingent outcome of the unfolding controversy. Under pressure, behavior geneticists sought to create space for maneuver. How to do this was a matter of dispute: a minority wanted to purify the field by expelling Jensen, but most sought to situate behavior genetics in a middle ground between Jensen and his opponents and to construe the controversy in ways that limited the scientific, political, and social tethers on their practice. Both efforts would fail. Years of increasingly polarized combat with critics would bind Jensen and his allies together with behavior genetics.

"Make a Virtue of the Middle Ground"

The young field was just in the process of forming when the IQ controversy struck, and its porous, ill-defined boundaries, still weak institutions, and lack of a strong identity facilitated Jensen's appropriation of its concepts and sowed ambivalence among behavior geneticists about the best way to respond. One view was that Jensen's work misused and misinterpreted behavior genetics, and the field should push him out of its bounds. Animal behavior geneticist Jerry Hirsch and several of his allies were the chief proponents. Early on Hirsch showed how several behavior genetic concepts undermine Jensen's central claim of an inverse relationship between heritability and educational improvement and warned that it was the fallacious use of genetic reasoning to justify racism (against blacks and Jews) that led to the rise of extreme behaviorist environmentalism in academia.[24] Thus Hirsch saw Jensen and allies as threatening the pact holding together behavior genetics. As the controversy unfolded, Hirsch became increasingly dismayed that behavior geneticists were doing little of the policing that would reestablish the pact. He delivered a series of talks that culminated in the publication of "Jensenism: The Bankruptcy of 'Science' without Scholarship,"[25] a savage takedown based on an extraordinary dissection of the scientific claims and discursive practices of Jensen, his allies, and the behavior geneticists he considered culpable in their success. For Hirsch and his allies, these actions had broken the behavior genetics pact, and they attacked with abandon not only Jensen's crowd but also much of the psychometrical branch of human behavior genetics.[26] They attacked as if the field had to razed to the ground before it could be reconstructed again on valid foundations.

However, most behavior geneticists saw the core of the crisis as the irrational, political, and sometimes-physical response that Jensen and they themselves were facing in what should be an intellectual debate. A crisis like this was precisely why an implicit pact for mutual support was necessary—after all, a crucial reason for the agreement was securing scientific autonomy for behavior genetics by freeing it from political manipulations. Further, while Jensen was certainly a provocative author both partly culpable for the attacks he was receiving and capable of defending himself, he was also a victim of anti-intellectual intolerance, and he could not be blamed for the flames spreading throughout the field. One now eminent psychological behavior geneticist recalled how dangerous the times felt for all behavior geneticists:

Then '69, Jensen's article and "boom" it just was dead. You know, people thought that really was going to be the end of human behavioral genetics, because the reaction was so severe. And that's when I was in graduate school,

1970, and I remember going to my first conference, which was the Eastern Psychological Association in Boston.[27] And Leon Kamin [who later wrote *The Science and Politics of IQ* (1974)] gave his first talk on, you know this rant against genetics—really ad hominem stuff. He's a great speaker and a real rabble-rouser. He had people on their feet. You know, you felt like there was going to be a witch hunt. It was my first meeting. And I was just an honest behavioral geneticist, you know I'm just interested in genetics and environment and the causes of behavior, and it was my first exposure to how violently enraged people can be against something which I consider an intellectual issue.[28]

Rather than dissolving the pact and turning on Jensen and each other, most behavior geneticists sought to act together to carve out an independent intellectual space for the field free from politicized accusations about it.

The proper approach, on this view, was to take a middle-ground position, taking a reflective stance on the substantive matters while defending researchers' freedom to work in this area without being harassed. The human behavior geneticist Irving Gottesman articulated this approach when he asked, "how do we shield the fledgling behavioral genetics both from the choking embrace of its friends and from the uninformed rejection by its antagonists?" He hoped the field would be able to avoid "the evils of both geneticism and environmentalism and then [make] a virtue of the middle ground."[29]

This middle-ground approach became the orthodoxy among behavior geneticists. While Hirsch lambasted Jensen and allies as pseudoscientists, most tried to approach their claims dispassionately as ordinary scientific ideas. For example, in an effort that included human and animal researchers, behavior geneticists organized a conference to evaluate the claims Jensen and others had made and clarify the field's contribution to educational policy.[30] During the conference, the lack of evidence for the genetic hypothesis for racial differences was reaffirmed. There was also an exchange in which behavior geneticist John DeFries explained to Jensen how he had misused one of DeFries's equations to infer between-group differences from the calculation of heritability within groups.[31] Researchers applied alternate statistical models to the available datasets to evaluate his 0.8 estimate of the heritability of IQ.[32] Behavior geneticists also produced several empirical tests of the genetic hypothesis for racial differences. For example, Sandra Scarr's study of adopted black children in Minnesota showed closer IQ correlations to their white parents than to their biological mothers.[33] She also studied racial "admixture" and found that black children with more white ancestry performed no better on tests.[34] Paul Nichols and V. Elving Anderson studied a multiracial cohort of children born at several urban hospitals and found that controlling for class largely closed the racial IQ gap.[35] Incidentally, no behavior geneticist produced empirical work that affirmed genetic evidence for the racial IQ gap.

Sometimes behavior geneticists' attempts to cast themselves in the middle ground were quite literal. In producing their book-length review of research on the causes of racial IQ differences, John Loehlin, Gardner Lindzey, and James Spuhler sent the manuscript to four sets of formal and informal advisors, sets that included prominent representatives of both sides of the controversy. Sandra Scarr's book *Race, Social Class, and Individual Differences in IQ*, which synthesized her decade of work as a psychological behavior geneticist, gave Jensen and Kamin space to articulate their views, and then Scarr positioned her behavior genetics work between them.[36] Even behavior geneticists like Scarr who were quite close to Jensen intellectually and apparently personally were being careful at this point to leave daylight between their positions and his.[37]

Importantly, behavior geneticists took on the crisis as a collective problem. Not only psychometric behavior geneticists whose data and methods were most directly involved participated, but many researchers whose work focused on animals and was not implicated in the debate also participated in the conference and other efforts.[38] And even as the controversy progressed and, as we will see, the field became scientifically boxed in with Jensen's ideas, behavior geneticists still vigorously maintained their independence and middle-ground position at a discursive level.

The External Critics

Behavior geneticists were not able to handle the controversy internally and were forced simultaneously to deal with external critics. Scientific issues were, of course, central, but the politics of the controversy and the terms of participation became crucial as well. On all these issues conflict with critics became increasingly polarized, the middle-ground position became unsustainable, and, as they tried to preserve intellectual and practical space for themselves, behavior geneticists were pushed and pulled into close association with Jensen and his allies.

Much of the early criticism of Jensen's ideas from those outside the field was compatible with members' efforts to claim the middle ground and distinguish themselves from Jensen. For example, James Crow, one *Moderate voices* of the first geneticists to respond to Jensen, noted that he mostly agreed with the analysis, but that he was more cautious about the models and thus would emphasize more than had Jensen that heritable traits can still be changed and genetic causes of racial gaps are extremely difficult to substantiate.[39] That heritability doesn't imply immutability nor does it yield information about racial differences were common critiques geneticists like Joshua Lederberg, Theodosius Dobzhansky, Walter Bodmer, and Luca Cavalli-Sforza made at this point.[40] Behavior geneticists said these things too, and some would later note that though he was rarely credited for it Jensen himself had included the proper technical qualifications.[41]

But a set of more trenchant critiques, which emerged at the same moment and soon drowned out these moderate voices, rejected behavior genetics' claim on the virtuous middle and, indeed, rejected that framing of the debate. Psychologist Leon Kamin examined in detail the twin and adoptee datasets that Jensen had used, arguing that flaws rendered all of them unreliable, and thus that no "prudent man" should accept the conclusion that IQ is heritable.[42] Kamin didn't merely attack Jensen's interpretations of behavior genetics data or the high estimates of heritability as had others, his critique suggested that the entire field lacked any empirical basis.[43] Kamin also highlighted problems in Cyril Burt's dataset of twins reared apart that suggested it was fraudulent. This was a shocking blow to the field's credibility, and behavior geneticists decried Kamin's "scorched earth" tactics.[44]

Rejected everything

Geneticist Richard Lewontin's first foray as a Jensen critic echoed the moderate critiques when he illustrated in detail why heritability was a local variable that didn't imply trait immutability, nor could it illuminate group differences. But he went further by accusing Jensen and other educational psychologists of being in the seventeenth century scientifically and blaming their own ignorance of how to educate children on children's genetic ineducability.[45] Later, he, Marcus Feldman, and other geneticists attacked the idea that heritability estimates could ever yield causal information, which would be necessary to build policies or inter-

ventions, about how genotypes produce phenotypes in response to environments.[46] These charges went much further than critiques of the re-
liability, assumptions, or interpretations of heritability estimates. They
argued that behavior genetics, in essence, had no data relevant to the
questions it aimed to answer, and, worse, that it had no prospects of getting data that would ever mean anything.[47] Thus the thrust of these critiques was to box Jensen and his allies together with behavior genetics
and to constrict—even to the point of nonexistence—the legitimate scientific space for behavior genetics, especially human behavior genetics.

Arguments about the politics of the controversy were also crucial in
polarizing the debate and driving behavior geneticists and Jensen together. The politics of the controversy have often been interpreted in
terms of a priori political commitments: Jensen and behavior geneticists as center/right and critics as liberal/radical.[48] But this view under-
plays the degree of liberal sentiment within behavior genetics and also
the ways that political affinities were often arrived at or ascribed as the
controversy unfolded.[49] One animal behavior geneticist, indeed a critic
of the association with Jensen, described this process:

> Within that field the participants became polarized. A group advocating the
> genetic control of behavior and a group saying that that couldn't possibly be
> true, that essentially phenotype is an interaction and you can't partition it
> into genes versus behavior [*sic* environment]. Right, that debate which was all
> within quantitative behavior genetics became polarized and I think to some
> degree there were people who were politically motivated that developed ar-
> guments on each side of the spectrum. But by the same degree, then, what
> happened is that people who argued one side or the other were immediately
> labeled as being politically liberal or politically conservative when in fact they
> were not motivated by any political persuasion, but rather were trying to stick
> with the scientific arguments.[50]

While left versus right politics have always been consequential for the interpretation of the controversy by observers as well as participants, this
understanding has been overplayed.

Far more crucial was the conflict over the relationship between science and politics. Behavior geneticists tried the scientific issues as separate from their political implications, which was an effort to preserve
space for behavior genetics as legitimate science. They argued that they
simply evaluated the data they saw objectively, that a heritability estimate

was just a number with no politics about it, and that their research didn't dictate what society chose to do. Thus while Jensen may have claimed that the high heritability of IQ shows that compensatory education must fail, other behavior geneticists could argue that it merely meant that bigger environmental changes had to be tried—but this was to make the argument about policy choices, not the integrity of the science.

Building on this frame, behavior geneticists thought they could establish common ground with critics by arguing that intellectual freedom as a common academic value was under threat when protestors protested or threatened scientists—whether Jensen, Herrnstein, Eysenck, or the behavior geneticists who would present their ideas for debate. For example, in 1973 Sandra Scarr invited Richard Lewontin to issue a joint statement denouncing recent claims that political racists had made about both their work on racial diversity while asserting scientists' academic freedom in this area.[51] But Lewontin sharply refused, responding: "the issue is not and never has been one of academic freedom. . . . The simple and direct fact is that genetics is being used as a weapon in a social battle and we disarm ourselves completely if we allow the battle to be fought on the ground laid out by racists."[52] Thus critics refused to credit Jensen and some others' work, perhaps even behavior genetics more broadly, as bona fide science that should be protected by the right to academic freedom. Further, they charged that the academic freedom debate was a smoke screen for the real issue of scientific racism, and they implied that its supporters, perhaps naively, were promoting racism.[53]

Further, critics rejected the idea that "political implications" (a phrase that already presumes a divide) were separate from knowledge in this area. They pointed out that many of their opponents were at best incautious in drawing sweeping political and policy implications from a few small twin studies, that the claims being made on behalf of behavior genetics were similar to those advanced earlier by eugenicists, and some claimed that errors in Jensen's and others' writings were biased toward findings with more conservative implications.[54] Others insisted that behavior genetics was inherently political regardless of scientists' personal motivations because the scientific relevance of its topics (for example, IQ, racial differences) was given by their political relevance, and, what's more, claims about their heritability only made sense if one accepted that the social order (that is, the environment) was given as is.[55] Further, radical critics claimed that behavior genetics was one arm of a larger revival of biological determinism together with theories like E. O. Wilson's sociobiology, which identified rape and xenophobic violence as naturally evolved human survival strategies and used hormone and brain research to explain gender inequality.[56] These accusations left no space for a non-ideological science of behavior genetics. Behavior geneticists may have disapproved of the habits of Jensen, Eysenck, Herrnstein, and Shockley to draw bold political implications from their science, but these, at least, could be assimilated in their framing of the politics/science relationship.

Despite Hirsch's effort to preserve behavior genetics by expelling Jensen and his allies, the mainstream approach of behavior geneticists was to manage the controversy by situating themselves in a middle ground to open up room for maneuver. Initially, critics seemed willing to accept this arrangement challenging discrete aspects of Jensen's claims. But as critics became more radical and moderates were crowded out, they aimed to box together Jensen and behavior genetics, charging their efforts to be scientifically impossible and thoroughly politicized. The effect was to close off the middle ground as a viable space for behavior genetics to occupy and for Jensen and his allies to become incorporated into the field—with enduring intellectual and social consequences for the field.

The Intellectual Stakes

The controversy had a profound impact on the intellectual organization of behavior genetics. By the end of it they would endorse three crucial points that Jensen had asserted or implied: first, the importance of making heritability estimates; second, a causal interpretation of heritability; and third, that genetic differences explain some of the black/white gap in IQ. These ideas, especially the first two, were not new to behavior geneticists, but that they should occupy a central role in organizing activity in the field was new. As I argued last chapter, behavior geneticists had hoped to move beyond the nature versus nurture debate, with its emphasis on partitioning population variance into environmental and genetic components, by de-emphasizing heritability studies, emphasizing their limitations, and clearing space for new approaches to emerge. But the IQ controversy got many behavior geneticists heavily invested in these types of research and strengthened the salience of the nature versus nurture framework. Race differences had been pushed even more forcefully off the agenda, but behavior geneticists ended up reversing their stance and endorsing Jensen's claims, which put them in an extreme position with respect to all their neighboring scientific fields. These intellectual stances were not due to preexisting political or scientific commitments, but rather how these commitments came to be redefined through the defense of the field.

These commitments were propelled by a set of intellectual standards that behavior geneticists deployed in sharp contrast to their critics' views. Although the middle ground ceased to be a viable position to occupy, it continued to be an orientation animating behavior geneticists' intellectual judgments, They pursued an evenhanded approach to data in which they tried to balance evidence from both sides and were tolerant of the problematic data in this area yet sensitive to the limits in their interpretation. For example, twins reared apart might have been separated at different times and to different degrees, IQ assessments often relied on varying instruments, and scores for biological parents (who had often given up their children and were impossible to locate) were sometimes estimated from caseworkers' impressions. The lack of clean, consistent data meant that behavior geneticists had to trust each other's authority both in their generation and in their careful interpretation.[57]

Critics, in contrast, had a much more absolute standard of evidence. Ambiguous, equivocal, or inconsistent data they saw not as suggestive, partial, or limited evidence but as nonevidence that should not be counted as science.[58] Behavior geneticists thought the accumulation of ambiguous data would gradually build a picture of certainty.[59] Critics held that errors accumulate rather than cancel each other out. Where behavior geneticists managed these ambiguities by adding qualifiers and pointing to uncertainty in their interpretations, critics accused them of issuing "cant"—dressing up qualified falsehoods as partial truths.[60]

These two sets of intellectual standards were partly linked to the differences between observational and experimental sciences. And Lewontin linked them to differences of political outlook—liberal, gentlemanly evenhandedness versus Marxist critique.[61] But they clearly congealed out of the intellectual combat of the IQ controversy. After all, behavior geneticists became more tolerant of data, especially about race, than they had been previously. And critics didn't turn their microscopic skepticism on other fields of research.[62] Yet what is crucial is that both these standards of evidence fit rational definitions of science despite the two parties' mutual accusations to the contrary. And their differences help us understand how the intellectual stakes of behavior genetics were established and how there could be persistent disagreement about them.

Heritability Estimates

One set of stakes that came to animate the field starting in this period concerned the tools and claims for estimating heritability. A key point in the IQ controversy was Jensen's claim that based on the results of different studies using various methodologies the heritability of IQ is about 0.8. Heritability is defined as the portion of the variance in a trait for a given population that is due to genetic variance rather than environmental variance. The 0.8 estimate suggested to Jensen and many others that genetic differences are about four times more important than environmental differences in explaining differences between individuals in IQ scores.

Leon Kamin methodically criticized each of the datasets that Jensen had used to make this estimate, and he argued that each had flaws that should lead researchers to abandon them as scientifically meaningful. Kamin's conclusion was that a "prudent man" had no warrant for rejecting the null hypothesis that the heritability of IQ is 0.[63]

These became two focal points of the debate and were often perceived as its extreme positions with the true value lying somewhere in between. The sociologist Christopher Jencks and his colleagues captured the spirit animating the field when they wrote that "the real question is not whether such [genetic] differences exist, but whether they are large or trivial."[64] As historian Diane Paul has pointed out, the fundamental ambiguity of this question itself helped spur the debate—what, after all, constitutes a "large" number here? "Is 0.50 a large number?—Is 0.60?—Is 0.40?—Is 0.25?"[65]

A tremendous amount of scientific activity propelled the debate.[66] Thus there were debates about whether twin studies generated estimates that were too high; for example, because identical twins experience more similar environments than fraternal twins. Or whether studies of adoptees were superior to studies of twins. Or whether natural human populations met the various technical conditions (such as random mating) for valid heritability estimation. Or whether it is possible to detect gene-by-environment correlations (for example, high IQ parents give their children both genes and more stimulating environments) or interactions (that is, that a particular genotype may influence IQ differently depending on environmental conditions) with particular research designs and, thus, whether they confound heritability estimates.[67]

Over the course of the debate an implicit standard emerged that higher numeric heritability estimates were "good for" behavior geneti- *↑ tiered* cists while lower ones were "good for" their critics. Thus behavior geneticists tended to favor relatively high heritability estimates, and field members could signal their status as moderates by backing a somewhat lower estimate. One such individual, a psychiatric geneticist who has been a member of the field since the 1970s, told me that behavior geneticists also tended to favor twin studies and statistical models that produce these higher estimates and to ignore ones that produce lower scores.[68] All of these factors set the terms for the field's continuation and expansion as the IQ controversy cooled late in the 1970s. Behavior geneticists often acted as if the point of the field was the accumulation of more and more heritability estimates for a wider range of psychological and behavioral traits. And this is in large part a legacy of ongoing controversy and skepticism whose origins were the zero heritability claims of the IQ controversy.

The Causal Interpretation of Heritability

One major tactic of behavior genetics' critics was to argue that heritability scores, whether their value was high or low, yield no information about the cause of a trait and therefore they are meaningless for making claims about whether that trait might be changed. The simple version of this argument was that heritability concerns the partitioning of variance around a mean, not the magnitude of the mean itself, and in educational policy, crime policy, and the like we care much more about increasing or decreasing means than manipulating population variance.[69] But the deeper point for many geneticists was that heritability scores are necessarily linked to a specific population and a specific environment, and they do not reveal the effects of changing the environment or the population.

To illustrate with an analogy, growing open-pollinated (genetically variable) corn in a shaded greenhouse would yield a high estimate for the heritability of height since the environment is very consistent, and most variation in height will be due to genetic variation. However, if the experiment is repeated, but the shading is removed from the greenhouse, the average height will be much higher, but the heritability will be similar.

Behavior geneticists accepted that it was true that high heritability did not mean a trait could not be changed through an environmental intervention, but they thought that if the heritability was high and the sample used to estimate it was diverse enough so that one could be fairly certain to have had a good distribution of genotypes and environments, then one could infer that heritability described how much the population could be expected to improve given the current range of environments. As Loehlin, Lindzey, and Spuhler argued:

> Most proposed policy changes involve minor redistributions of environments within the existing range, and it is precisely regarding such changes that a heritability estimate has its maximum predictive value. For instance, one message that a high heritability coefficient can convey is that minor fiddling around with environmental factors that already vary widely within the population has poor odds of paying off in phenotypic change—and thus new ideas about environments need to be tried.[70]

For behavior geneticists, the ~0.8 heritability of IQ suggested that gains would not be achieved by the educational equivalent of moving to a better part of the greenhouse or five more minutes of sunshine. But major reforms—de-shading the greenhouse—would be costly and have unknown effects.

Critics probed heritability further by discussing norms of reaction.[71] Norms of reaction represent the phenotypic value (say of corn height or IQ scores) of particular genotypes that develop under particular environmental conditions. Figure 3.1, which reproduces Feldman and Lewontin's diagram, gives several hypothetical examples. The best way to generate such data for humans would be to assign children with different genotypes to grow up in different environments. For better or worse such an intervention is ethically and practically impossible. Many have hoped that heritability scores, which offer a global representation of the sources of variance of an underlying structure like this, could be used to infer backward to the causes. The problem, critics argued, is that the heritability score is highly dependent on the shape of the reaction norms, how common in the population are particular genotypes and environments, and how genotypes are distributed into environments. Behavior genetic studies of twins and adoptees may control genetic and environmental variance in certain ways, but they do not yield information about the underlying shape of the curves or where in the genotypic or environmental distribution the sample comes from. Heritability scores cannot on their own help construct the reaction norms, but without a reaction norm the causal relationship between genotype and environment cannot be unpacked.

Further, if figure 3.1 represented the norm of reaction for IQ in different educational regimes, then substantial gains for the population could be made by ensuring that G2 kids were in environments on the left and G1 kids on the right rather than everyone being randomly distributed. Lewontin criticized the "minor fiddling around" point by saying that it confused the *range of environments* issue with the *distribution of environments*, and that "what we require first is not new methods of schooling but an equitable redistribution of extent resources."[72] Behavior geneticists again did not dispute the technical points, but they thought the norms of reaction for IQ looked something like figure 3.2. Everyone does worse in "bad environments"—poverty, bad schools, uneducated parents—and better in good ones. Even in good environments there is still variation, and some kids will do better than others. Maybe black children were mostly on the Genotype A curve, which would account for their often lower than expected performance in middle-class environments. Maybe Jensen was right about differences in educability.

Here again the two standards of evidence came into play. Critics held that genuine norms of reaction were basically impossible to get for humans because they cannot be experimentally randomized into developmental environments. Experimental evidence from plants and animals suggest that the shapes of the curves cannot be inferred in advance and rarely follow the smooth, nonintersecting pattern like figure 3.2.[73] Thus true causal interpretations of heritability are hopeless and must be abandoned. Behavior geneticists did not claim direct experimental evidence, but they thought these various indirect lines of evidence provided a reasonable set of assumptions that would enable them to interpret heritability scores causally—provided they offer appropriate, reasonable qualifications.

Genetic Causes of Racial Differences

Over the course of the controversy, two basic positions emerged on the genetic hypothesis for the IQ gap between blacks and whites. The first

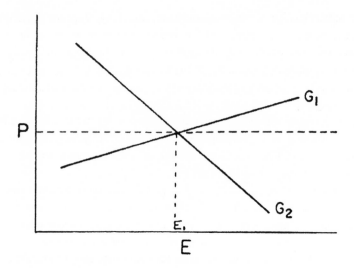

FIGURE 3.1. Norms of reaction from Feldman and Lewontin (1975, 1166)
Note: Phenotype P plotted against an environmental variable E. If the environments are symmetrically distributed around E_1 there is no average effect of genotype. If there is an excess of G_1 in the population the average phenotype will be constant, as represented by the horizontal dashed line.

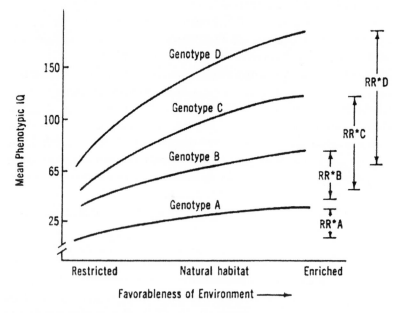

FIGURE 3.2. Norms of reaction from Gottesman (1968, 33)
Note: Scheme of the reaction-range (RR*) concept showing the interaction of heredity and environment. (From Handbook of Mental Deficiency, edited by N. Ellis. Copyright © 1963 McGraw-Hill, Inc. Used by permission of McGraw-Hill Book Company)

was Jensen's claim, echoed by Eysenck, Shockley, Vernon, Cattell, and others, that a preponderance of imperfect evidence "strongly implicated" genetic factors in explaining the gap. The second, articulated by Lewontin, Hirsch, Feldman, Bodmer, and Cavalli-Sforza, was that the evidence was decidedly inconclusive. Even Jensen would admit that the decisive genetic experiment—which he said would involve crossbreeding blacks and whites and randomly cross-fostering the children—is ethically impossible and "the problem taxes scientific ingenuity, and is hence probably insoluble."[74] And his opponents did not claim to know that genes had no effect on the IQ gap.[75]

The disagreement, once again, was whether or not circumstantial, indirect, inconsistent evidence could be counted as positive evidence. Jensen and his allies said *yes* while critics said *no*. As I showed last chapter, prior to Jensen's efforts, prominent behavior geneticists claimed that there was no good evidence for genetic causes of racial differences in intelligence or other important behaviors, that there was no prospect for getting good evidence, and that the topic should be dropped.[76] This view, basically the same as the critics', was the uncontested orthodoxy, if not the consensus, of behavior geneticists.

In 1971 a group of behavior geneticists decided to take up the task that Shockley had proposed and the National Academy of Sciences had declined by systematically reviewing the evidence for genetic and environmental causes of race differences in intelligence. The group members were psychologists John Loehlin and Gardner Lindzey and physical anthropologist James Spuhler—the latter two had previously written one of the decisive reviews pushing the topic of race differences off the behavior genetics agenda.[77] They were supported for a year of work by the Social Science Research Council, the Center for Advanced Study in the Behavioral Sciences, and the US Office of Child Development and were advised formally and informally by dozens of scholars.[78] The aim was to produce something like a decisive, quasi-official consensus statement whose reasonableness and rationality would contrast with so much of the way the IQ debate had been conducted.

The result, published in 1975, was *Race Differences in Intelligence*, a 380-page monograph with fifteen appendixes and hundreds of references. The crucial passage summarizing the etiology of racial differences was: "Observed average differences . . . probably reflect in part inadequacies and biases in the tests themselves, in part differences in environmental conditions among the groups, and in part genetic differences among the groups."[79] They reached very much the same conclusion that Jensen had, though perhaps with slightly less emphasis on genetic causes than his "strongly implicated" statement. This conclusion was not only a reversal of the previous orthodoxy in the field, it was reached despite the fact that the empirical evidence that behavior geneticists had produced in the interim did not support the genetic hypothesis.

Loehlin, Lindzey, and Spuhler's endorsement of the genetic hypothesis was a product of their efforts to be reasonable, judicious, respectful, and fair to all of the different ideas and data scholars had brought to the issue—it was a product, in other words, of their middle-ground ethos. Behavior geneticists and their allies seem to have interpreted this effort as marking the field's distinctively balanced, dispassionate, and intellectually courageous approach to a problem that was generally emotionally and politically charged. If behavior geneticists disagreed with the framing or conclusions, they did not publicly criticize the work.

Critics, in contrast, thought the balanced and respectful approach to all the ideas meant that the book encompassed and reproduced the contradictions of the literature rather than cutting through them. Lewontin argued that for all the effort they expended, only a few pages of evidence on interracial marriages bore directly on the genetic hypothesis for racial differences, and including everything else only muddied the picture.[80] And these few pages provided no evidence for the genetic superiority of whites. He wrote:

The failure of *Race Differences in Intelligence* to provide a hard and incisive analysis of the problem it sets itself arises from the belief structure of the scientific community which, in turn, reflects one of the guiding unexamined principles of intellectual life. It is the principle that the truth about anything always lies about halfway between the most extreme possible positions . . .

Words like "absolute," "zero," "always," and "never" are anathema to the tradition that gives rise to this book. "Relative," "not significant," "usually," and "sometimes" are the more comfortable, less threatening rhetoric we find in it. But the real world is different. One plus one is never two and a half. Our world of intellect is a world of two-valued logic; if A does not logically entail B, we are not allowed by some muddleheaded, middle-of-the-road ideological commitment to suggest that it sort of does, at least in months with an R in them.[81]

With characteristic sarcasm Lewontin explicitly attacked behavior geneticists' middle-ground approach to the controversy, arguing that their
evenhandedness made them accept false evidence.

To the extent that *Race Differences in Intelligence* represented the
orthodox view of the question in the field, it situated behavior genetics
as an outlier on the genetic hypothesis relative to other fields. In 1975 a
group of geneticists, dismayed by the IQ controversy, mobilized to get
members of the Genetics Society of America (the main scientific society for geneticists) to pass a resolution against the hereditarian interpretation of IQ and racial differences. Over nine hundred of the society's
2,600 members attached their names to a statement that read, "there is
NO CONVINCING EVIDENCE OF GENETIC DIFFERENCE IN INTELLIGENCE BE-
TWEEN RACES." And 1,390 agreed to associate their names with a subsequent revised statement that read, "In our views, there is no convincing
evidence as to whether there is or is not an appreciable genetic difference in intelligence between races."[82] Although Loehlin, Lindzey, and
Spuhler had stated support for the genetic hypothesis—with the words,
"probably reflect in part"—in a way that seemed cautious and reasonable, substantively it was radical compared to the genetics profession as
well as many professional social science associations that had produced
similar statements critical of Jensen's claims.[83]

Although this position set the field apart from its neighbors, it did
not lead to a "mainstreaming" of race differences as a research topic in
behavior genetics. Whatever their views on the reality of genetic racial
differences or the adequacy of the preponderance of imperfect evidence
standard, it was clear that this was still a difficult, unattractive, and professionally dangerous topic that most preferred to ignore. Instead of
spawning research, behavior genetics' position on this issue turned the
field into a kind of sanctuary, though not necessarily a warm home, for
the few researchers who did devote substantial attention to it—Jensen
and Eysenck at the time and later Herrnstein (when he turned to race),
J. Philippe Rushton, Richard Lynn, and Linda Gottfredson, for example. Behavior geneticists I spoke with are now uncomfortable with the
field's association with race research, but they tolerate race researchers'
right to the topic as a matter of intellectual freedom. Race researchers
sometimes complain about the lack of support from their colleagues, but
apart from a small number of critics within the field, they at least haven't
had to worry about being attacked.[84] Behavior geneticists' failure to fol-

low other fields and expel the topic has lent a persistent aura of scandal to the field. But this implicit agreement would enable the race controversy to cool off after the late 1970s, and it wouldn't become a huge issue again for the field until the mid-1990s with the publication of *The Bell Curve*.

Disconnecting from the Field of Genetics

The IQ controversy raised all kinds of questions about who should be legitimately able to mobilize genetic arguments, explain racial differences, and analyze educational problems. If this was a struggle for authority between psychology and genetics, one of the interesting features is that neither disciplinary core was particularly enthusiastic about occupying the intellectual space. As I showed in the previous chapter, behavior geneticists had strongly argued that psychology and the social sciences generally would be handicapped until they incorporated information about genetic heterogeneity. This argument in the abstract is one thing, but Jensen had raised the stakes by attributing the specific failures of psychologists to educate children and reduce racial inequalities to their ignorance of genetics. Since psychologists were heavily invested intellectually, practically, and reputationally, in compensatory education efforts, it is not surprising that they would fight against the entrée of behavior genetics into this realm.[85] Geneticists on the other hand had no problem with the idea that genetic causes affect behavior, but they worried that the IQ controversy was reviving the stigma of eugenics and racism from which they had long worked to cleanse themselves.[86] Of course, members of these disciplines were not unified in their perceptions of the value of this space—the existence of a thing called *behavior genetics* and a controversy about it is predicated on deep differences within disciplines on these matters. However, from the perspective of the disciplinary cores, it was more like a struggle for authority in reverse: rather than fighting for the right to occupy this intellectual space, as the controversy got uglier they could be seen as fighting to keep the space unoccupied.

But the capacity of the traditional disciplines to "govern" behavior genetics was limited. First, while it might be a bit overstated to call those interested in participating in behavior genetics "dissidents" from their home fields, by virtue of investing their efforts in behavior genetics, which sought to change those disciplines, they had already to a degree shed some of the normative binds of the disciplinary centers. Second, and closely related, behavior genetics' nascent institutions, chiefly the Behavior Genetics Association and the journal *Behavior Genetics*, were new and small, but they were sufficient to provide an alternative outlet for these quasi-dissident scholars. And third, the highly statistical character of much research in this area meant that the barriers to participation were quite low for those with a certain skill set. The words of Arthur Goldberger, an economist and skeptic about the too easy use of genetics in social science, are telling: "At first I had thought, naively, that that analysis was genetics in the sense of microscopic examinations. Then I realized that the data consisted of correlation coefficients. . . . That was easier for me, I didn't have to learn any biology in order to go on."[87] Many on both sides of the controversy had no prior training in behavior genetics but used their statistical expertise to participate. After all, modern statistical methods used in all fields were invented by biometrical geneticists often to analyze behavioral traits, so this was in some sense the chickens coming home to roost.

The efforts of population geneticists like Lewontin, Feldman, Bodmer, and Cavalli-Sforza to point out the many problems with heritability estimation from twins and adoptees can be seen as symbolic boundary work against statistics-oriented psychologists to claim they lacked the competence for authentic genetic analysis. But in claiming this research was impossible to do with humans and even difficult with animals and plants, they aimed to warn away all scientists. Rather than directly attacking psychologists' genetics competence, critics reversed tactics on them by challenging the psychologists' authority over the ostensibly psychological topics.[88] Thus nonpsychologist critics frequently attacked the concept of IQ, the data about educating children, and even the accuracy of the psychologists' representation of psychological research.[89]

None of these salvos convinced Jensen, his allies, or behavior geneticists to leave the field (at least not those active in the controversy). Instead, their major impact was to dissuade geneticists from pursuing research in this area. There were certainly geneticists sympathetic with what Jensen had argued and with the broader project of behavior genetics. Jensen's gentler critics seemed to accept that the simplifications involved in estimating heritability did not imply the task was meaningless.[90] And many eminent geneticists and biologists—among them Dobzhansky, Francis Crick, and Ernst Mayr—were sympathetic and even enthusiastic about the agenda Jensen had espoused and were quietly angry with the attacks of Lewontin in particular.[91] Indeed, Crick proposed starting up a "twins institute" to bolster the limited collection of twins separated at birth—an idea that would prefigure the famous Minnesota Study of Twins Reared Apart launched by behavior geneticist Thomas Bouchard.[92] Mayr considered the research crucial for the eventual implementation of nonracial, eugenic social policies he considered necessary.[93]

But the crucial point is that whatever their sympathies, geneticists largely heeded the population geneticists' warnings. The GSA's collective statement against the genetic hypothesis for race differences is one sign. No major geneticist publicly took on the critical population geneticists or backed Jensen. Geneticists who did sympathize did not turn their research toward behavior genetic studies of psychological traits in human populations. They were also less interested in joining the Behavior Genetics Association than in the past. A survey administered to the membership about their training showed geneticists only made up about a quarter of nonstudent members in 1979. But the students who were joining the field had emphasized coursework in the behavioral over the biological sciences more than their older colleagues. This suggests psychologists were joining the field in greater numbers than geneticists.[94]

And indeed Lewontin and Cavalli-Sforza were able to institutionalize their critique of behavior genetics by blocking the NIH funding of heritability studies. They were members of the Genetics Study Section of the Division of Research Grants at the NIH, which in September 1974 held a workshop to address the many issues of disagreement in population genetics; one of the few points of consensus the nineteen members reached was that "simple studies of heritability of human traits were worthless. . . . Moreover, it was agreed that heritability studies of normal human variation [as opposed to pathology] were misleading, especially human behavioral traits."[95] Since this group was charged with the review of all NIH grant proposals in population genetics, this statement ostensibly closed an important door on funding the kind of research that behavior genetics was emphasizing at the time.

Geneticists would not become seriously interested again in behavior, especially human behavior, until molecular genetic techniques developed to the point that it would become possible to link behavioral traits to genetic differences at the level of DNA. Occasionally, it seemed possible to make such claims in the 1970s and earlier. For example, geneticists long knew of particular mutations that led to forms of mental retardation (for example, Down syndrome, PKU, Fragile X), and there was a good deal of excitement about claims that an extra Y chromosome was responsible for criminal behavior.[96] However, it was really not until the early 1990s when the Human Genome Project was beginning—and promoted in part by the light it would shed on behavior—that geneticists returned to the topic en masse.

Conclusion

Behavior genetics had been founded on the conviction that it would be possible to launch a collective inquiry into the heredity of behavior without becoming dragged into the shameful politics of racism and coercive eugenics that had discredited previous efforts. Just as the young field was assembling its institutions and membership, the actions of Shockley and Jensen forced behavior geneticists to confront the problem of racial differences in intelligence on highly unfavorable terms. Media attention, protests, and academic criticism closed in on behavior genetics from all directions. Behavior geneticists struggled to maintain room for maneuver and the capacity to define the field on their own terms. Some felt the space could be maintained by summarily expelling Jensen and his concerns, but most believed it necessary to situate the field in the middle ground between Jensen and his critics. However, critics fought to close down the space for behavior genetics by strongly identifying Jensen with it and showing the whole enterprise to be scientifically and politically illegitimate. Under this onslaught, behavior geneticists found it necessary to defend and incorporate Jensen's position in order to defend the field.

The ways behavior geneticists defended themselves against critics would profoundly affect the field's intellectual preoccupations. Behavior geneticists based their authority on positive estimates of the heritability of intelligence and other traits; on a causal, generalizable interpretation of heritability estimates, and on "reasonableness" about the hypothesis of genetic explanation for racial differences in intelligence. This committed behavior genetics, by the mid-1970s, practically to the accumulation of heritability estimates and substantively to "hereditarian" interpretations of heritability and race differences, which is to say the perpetuation of the nature versus nurture debate.

By the late 1970s the terms in which behavior geneticists had previously imagined the field were fundamentally altered, and the slate of interests it made sense to pursue were redefined. The solidaristic ties of the "pact" had been strengthened, at least for those members who had not abandoned the field, but the substantive focus had shifted drastically. Behavior geneticists had hoped to depoliticize the field, setting aside race and eugenics, and moving beyond the nature versus nurture debate. However, they ended up endorsing not only the reasonableness of race research but also a version of Jensen's substantive claim about blacks' intellectual genetic inferiority—a sharp reversal of the collective position of the 1960s. Further, they became deeply invested in the nature versus nurture, heritability estimation framing of the field. Behavior geneticists' authority became wrapped up in generating more and higher estimates of heritability for intelligence and other behavioral traits.

The aspiration that the field would host scientists interested in the genetic influences on behavior in all its forms and feature the full range of experimental and theoretical positions on these questions had failed. Instead, behavior genetics became monopolized by a small set of the possible intellectual and practical positions. As we will track in upcoming chapters, rather than the "universal" field devoted to the genetic inheritance of behavior, behavior genetics became one of a patchwork or archipelago of largely disconnected subfields devoted to versions of the problem. Put differently, if a behavior geneticist from 1960 were rocketed forward in time by twenty years, he or she would probably have been surprised that so much of the field was involved in racking up heritability estimates, the higher the better. Further, it might have been surprising that scientists perceived behavior genetics to be associated with particular genetic methodologies and models and their critics with others, rather than having "behavior genetics" be a label that encompassed a debate about the merits of all of these elements. In other words, "behavior genetics" had become more like a set of positions in a broader space of intellectual contention rather than the field that housed the full range of contending points of view

The explanation I have offered of behavior genetics' association with Jensen and his ideas, and the rest of these outcomes, differs substantially from those usually offered. These weren't due to the preformed motives of behavior geneticists or the inherent structure of the field's ideas. Instead, the outcome was the product of a set of genuine dilemmas the field's members faced about how to defend the scientific space in the face of the IQ controversy. As the controversy unfolded and reshaped the field's practical logic, behavior geneticists' interests came to be redefined

in terms of preserving the scientific, political, and social room for maneuver. There were different conceptions of how best to do this, but behavior geneticists made these moves under serious immediate pressure
and for the most part did not anticipate or intend how they would come
to shape the field. They did, of course, adapt, and we turn now to these
adaptations in the context of other controversies.

Animals or Humans to Study Behavior?

Conflict over the Shape of the Field

Is research on humans or nonhuman animals better suited to help us understand how genes affect behavior? This question has underwritten one of the longest-running conflicts in behavior genetics. The advantages of animals—be they mice, fruit flies, nematode worms, or dogs—mainly concern the possibilities for controlling them experimentally.[1] Researchers can control animals' genetics via breeding, manipulate the environments in which they develop, set up precise conditions to measure their behaviors, and exploit their shorter life cycles to repeat the experiments. The key question is how to connect this research to humans. Should we think of the way a mouse crouches and twitches its tail prior to attacking another mouse as analogous to aggression in people?[2] When a mutated male fruit fly sexually mounts other males, is that homosexuality?[3]

The advantages and disadvantages of doing behavior genetics on humans are basically inverted. It is neither ethical nor feasible to control experimentally the genetic or environmental circumstances of human subjects.[4] Instead, human researchers look for "natural experiments" like monozygotic and dizygotic twins who grow up in different circumstances, or for subpopulations with high rates of some trait and high degrees of intermarriage, or well-known multigenerational family genealogies. Notwithstanding the difficulties in defining and measuring human behaviors, human researchers are relatively confident that the behaviors they study are the real things that matter in human experience. Much more problematic and contested in human research are the indirect tools that stand in for experimental manipulation of genes and environment.

The debate among behavior geneticists about the relative merits and problems of animal and human research and how the two should fit together was much more than an intellectual dispute. It was also a conflict over what kind of research would be most recognized and who would be able to impose their vision of behavior genetics. In Bourdieu's terms this was a struggle to define the field's scientific capital—which kinds of research would garner the most recognition and power for scientists conducting it. In the years following the IQ and race controversy the intellectual and social answers to the question were settled in a way that favored human researchers.

As I showed in chapter 2, behavior geneticists saw animal and human research as highly complementary in the field's early years. Humans *are* animals, so analyzing their similarities and differences should be revealing. Human behavioral problems would justify and motivate animal research while animal research would offer a model of noncontroversial scientific rigor for human studies. Part of this would involve human researchers adopting the "pure" orientation of animal researchers, carefully avoiding politicized topics, and studying behaviors closer to physiology than culture. It would also involve putting different species and different groups of humans in a comparative and evolutionary framework. Behavior geneticists also hoped that creative ways of designing human studies, drawing in part from animal research, would help human behavior genetics move beyond the nature versus nurture framing. Importantly, this wasn't only the view of animal researchers. Many human researchers were nervous about the shortcomings of their science, and many agreed that animal research held crucial tools for them. These intellectual arrangements were to be secured by a certain social order. Animal researchers took the lead in the intellectual framing and the institutional organization of the field. Their social vision was to establish the field as a robust transdisciplinary space where scientists from different backgrounds, interested in different questions, would engage and contribute to each other's projects. More than simply mixing scientists from different backgrounds, "behavior genetics" would be a new, hybrid intellectual space and scientific identity. It would have its own form of scientific capital and all members would prioritize the competition for it.

But that was just one way to answer the question in the opening. This chapter shows how human research became the dominant way to do behavior genetics. The terms of the pact that had previously bound together behavior geneticists were altered. Animal research was no longer seen as a model; indeed, behavior geneticists came to see the two sides as offering very little scientific value to each other. Animal researchers often tried to put their work in the service of human research, but were offered very little attention. Many animal researchers remained in the field, but they could no longer impose their vision. Human researchers had a different vision. They saw the point of behavior genetics mainly as allowing them to compete in their home disciplines rather than as engaging others in a hybrid interdisciplinary enterprise. Human researchers highlighted the evolved and animal nature of humans to attack the supposed biological denialism of social scientists, but not to generate a truly integrated research program. Practically speaking behavior geneticists stopped considering humans to be animals for the purposes of research. The newly dominant human researchers were not interested in building institutions, extending the network, or integrating the field. Instead of the site of a robust interdisciplinary conversation, they treated behavior genetics as a safe haven or bunker for embattled refugees of the IQ controversy. They saw the competition for behavior genetics' scientific capital as a secondary pursuit and a means for competing in other fields.

Unlike the IQ controversy, which was obvious to everyone, the conflict between human and animal researchers was subtle, not always recognized as such by its disputants, and entangled in other lines of conflict. This chapter shows how it was connected with transitions in the field's disciplinary makeup, as a certain strand of human clinical psychological research became ascendant over experimental psychology and other disciplines where animal research reigned. Behavior genetics' institutionalization in the university came into question as researchers concentrated in a small number of friendly departments. Additionally, practices of criticism and peer review came into dispute. Sociologist Robert Merton wrote about organized skepticism as an essential feature of science.[5] However, dominant behavior geneticists became skeptical about skepticism and suppressed scientific criticism and dissent that they considered scientifically counterproductive. These became sites for the conflict between the alternate visions of behavior genetics that human and animal researchers advanced. The chapter tracks how the vision of human researchers prevailed, and how, as a result, behavior genetics assumed the paradoxical archipelagic form I described in the first chapter.

Why did human researchers and their vision come to prevail in be-havior genetics? The turning point was the period after the IQ and race controversy when behavior genetics was in a vulnerable and unsettled state. In this moment of reconstruction, the kinds of critical conversa-tions necessary to the animal researchers' vision were difficult to have without being interpreted as disruptive and anti-behavior-genetics. Fur-ther, the recent controversy made the field more appealing to certain hu-man psychologists but less appealing to the multidisciplinary audience crucial to the animal researchers' vision. Many animal researchers left; those who remained were cut off from crucial allies. By the early 1990s, when the advent of molecular genetic technologies newly disrupted the field, human research had become dominant. Bourdieu highlights the eternal conflict between those seeking to preserve a scientific field's pre-vailing order and those seeking to overturn that order by redefining the field's capital. This chapter tells the story of the revolution that enabled human researchers to become dominant. Ironically, however, this was due less to a self-conscious revolution than to the collective adaptation, by both human and animal researchers, to the practical dilemmas of the field's unsettled state after the IQ controversy.

The story I tell of human researchers' ascendance is surprising in sev-eral ways. First, it contradicts the common notion that animal and hu-man researchers are working with incompatible "experimental systems," and that the controversy would be determined by these systems' capaci-ties and the questions asked.[6] I show that the causation is the opposite: different visions of the field imagine different relationships between ex-perimental systems and deem different questions worth asking. Second, it contradicts those who see the conflicts in behavior genetics as funda-mentally political.[7] The conflict between human and animal researchers *is* as much political as scientific, but it involves the social politics of the field much more than the broader politics of society. Finally, within sci-ence we might expect those higher in the prestige hierarchy to prevail in such controversies. Yet the human researchers mostly came from the lower-status social science end of the behavioral sciences while the ani-mal researchers tended to be aligned with higher-status biological and genetic fields. In contrast to all these other ways of viewing the contro-versy, my account emphasizes the importance of the specific social dy-namics internal to the field.

Disciplinary Reorganization or "There's a Lot of Behavior Genetics Going on, and There Has Been a Lot That Ain't Called Behavior Genetics"

The IQ controversy elevated the analysis of heritability and the issues of IQ and race, which turned off many geneticists, biologists, and social scientists who had previously been interested in behavior genetics. Its aftermath set in motion a disciplinary reorganization of behavior genetics at several different scales. The first of these involved the fragmentation of the integrated, interdisciplinary space the field's founders had aimed to create into an archipelago of different intellectual communities oriented toward different disciplines and research subjects (animals and humans). One psychiatric geneticist gave a keen sense of the field's archipelagic character in his description:

> So these are different clubs and people have different union cards. And the psychologists actually have one union card, the molecular geneticists have another union card, quantitative geneticists have a different union card, psychiatrists have a different union card. Now each one may overlap to some degree, but in fact these are people with different skills and interests and what you see is that these societies function as social clubs. And that is their social network, and you can belong to multiple clubs and you can be respected in multiple clubs, but as a psychiatrist, I'm much more accepted in the International Society of Psychiatric Genetics than I am in [the] Behavior Genetics [Association] even though, you know, for a time I was very active in that group.[8]

The speaker described behavior genetics here as a field fragmented mostly along disciplinary lines. The different islands each have their own practices and interests and have often somewhat tense relationships to other islands.

Another speaker, an animal researcher, put an even sharper point on the social and intellectual fragmentation of behavior genetics:

> What should make behavior genetics interesting [to a sociologist] is that it is not a field. It's a bunch of different groups of people who are trying to study different questions. Many of whom have completely insulated themselves against the outside world and scientists in other fields. They actually don't

want to know what other people say about them. Because they're so sure of the value of what they're doing and they're so sure that they're right that they've completely cut themselves off. And as I've sort of learned more about all of these communities, this to me is the most interesting thing. Because it seems to me to be completely antithetical to how you'd want to do good objective science.[9]

According to this speaker, behavior genetics has become so fragmented, and communication so stunted among its various islands, that he does not see it as a bona fide field. He invokes the field's bunkerization (which I discuss below) in the context of social fragmentation—the defensiveness of some has led the methodological and disciplinary differences featured in all multidisciplinary spaces to become hard, seemingly impenetrable barriers. It is noteworthy that both speakers, the psychiatrist who studies humans and the animal researcher, see the field as fragmented. The differing severity of their assessments is linked to the fact that human researchers tend not to find that the distinctions among them lead to an inability to communicate, but animal behavior geneticists find themselves cut off from animal researchers on other islands whose work they believe to be complementary.

Scholars have distinguished among different kinds of interdisciplinary fields.[10] An "interdisciplinary" field is one where researchers from different disciplines work jointly on a common problem. In a "multidisciplinary" field they work on a common problem but separately, in parallel or sequentially. And in a "transdisciplinary" field, they work together, blending disciplinary perspectives to create a new synthetic conceptual framework. What was happening in behavior genetics was the transdiscipline the field's founders (led by animal researchers) had attempted to construct was in this period unwinding into two directions. First, among human researchers it was becoming an interdiscipline where psychologists, psychiatrists, and a small number from other behavioral fields engaged a common set of tools and questions. Second, this interdiscipline became one cluster of islands within a broader archipelagic multidiscipline with islands of animal researchers with different degrees of identification and engagement with the field. The current state is what figure 1.1 represents.

A second order of disciplinary reorganization, then, involved differences in the degree to which members of different islands in the archipelago felt connected to each other or to the label of "behavior genetics."

Here is where the distinction between the field of "behavior genetics" and the field devoted to the "genetic inheritance of behavior" is particularly important. The "genetic inheritance of behavior" as a topical domain might be seen as analogous to the geographic proximity of islands in an archipelago or even the geologic forces that generate their grouping. But it is another matter whether island occupants identify with the label "behavior genetics"—which is akin to the political groupings that link some islands but not others. The field's founders saw no distinction between the two. They wanted "behavior genetics" to represent all the different researchers and approaches aiming to link genetics and behavior. But starting in the mid-1970s "behavior genetics" became much narrower practically speaking, and many researchers who fit the implicit "genetic inheritance of behavior" definition refused to identify with "behavior genetics." As one animal behavior geneticist put it to me: "there's a lot of behavior genetics going on, and there has been a lot that ain't called behavior genetics." Another animal behavior geneticist described his frustration at the unwillingness of many such researchers to engage behavior genetics:

> You have nowadays all kinds of people that are doing what I call behavior genetics, but don't tell them, because they don't regard it as behavior genetics. All these people doing knockout studies, and transgenic animals and things like that. They're looking at genes and the effect of genes on behavior. Well, how [else] should I define behavior genetics? But they don't see themselves as behavior geneticists. They hardly if ever sent their work to the journal *Behavior Genetics*. They never came to the BGA, although I tried to get them there, when I was still very active in the BGA. And the reason for that is that these people are neuroscientists. They're neuroscience. They don't even see themselves as geneticists. Least of all behavior geneticists . . . they are interested in brain mechanisms. . . . They want to know how is this working, what's going on?[11]

According to this speaker and others, many researchers from neuroscience, genetics, and other biology-oriented disciplines who use animal subjects to understand the ways that genes affect behavior through development, the construction of the nervous system, or brain physiology don't see their work as linked to the field of behavior genetics. For many of the animal researchers who do consider themselves behavior geneticists, this refusal has been a major source of frustration because they are partly cut off from these potentially valuable interlocutors.

For human researchers the stakes of the field's fragmentation were much lower. The divisions did not fall along major methodological or conceptual divides, nor did they generate major barriers to recognition or scientific practice, as they did for animal researchers. Rather, divisions concerned points of pride and the narcissism of small differences. For example, when beginning this project, I was advised by people familiar with the field not to approach potential interviewees from psychiatry asking to speak to me about "behavior genetics." They might be put off, I was told, unless I called it "psychiatric and behavior genetics." Psychiatrists and psychologists often mark the distinction between the domains of psychiatric genetics and behavior genetics by saying the former concerns illness or pathological behavior while the latter concerns behavior in the normal range.[12] However, in practice this distinction is muddled. One of the largest areas in psychology concerns the study of psychopathology, and clinical psychology is focused on understanding and treating psychological dysfunction. These foci have brought psychologists into jurisdictional conflict with psychiatrists.[13] Furthermore, some of the best-regarded early work on the genetics of schizophrenia—the mental illness par excellence—was done by psychologists Irving Gottesman and James Shields.[14] On the other side, psychiatrists like Robert Cloninger and Theodore Reich did important work on IQ and personality, the psychologists' favorite traits.[15] Psychological and psychiatric geneticists frequently span the disciplinary divide in research collaborations, citations, and edited collections, and they have used all the same methodological tools. Thus human researchers are cognizant of the different "islands" they occupy, but the differences are largely notional and have not been profound barriers to inter-island exchange.

A third level of disciplinary reorganization involved the increasing association of behavior genetics with the behavioral sciences and weakening ties with the biological sciences. Part of this shift was due to the efforts of critical population geneticists, which I discussed in the last chapter, to draw sharp symbolic boundaries around behavior genetics to define it as scientifically untenable and politically dangerous. This effort (combined with other factors at the time) was effective at stigmatizing behavior genetics and warning geneticists off the terrain. Relatively few geneticists were interested in joining behavior genetics at this point—not even, as I showed, those who were sympathetic with behavior genetics and irritated by the critics.

This realignment away from genetics helped precipitate a demographic shift among those newly joining the field's ranks. There were many geneticists and different kinds of biologists among the field's founders. But young behavior geneticists completing their training in the late 1970s tended to have more training in the behavioral sciences—psychology in particular—than their older colleagues. In 1979 the BGA surveyed its members about their graduate training in order to compare the education of older "Regular and Emeritus Members" to younger pre-PhD "Associate Members."[16] As table 4.1 shows, the median newly minted behavior geneticist had one fewer course in the biological sciences and two more in the behavioral sciences than older behavior geneticists. Further, the same proportion (16 percent) of Associate and Regular/Emeritus members reported no coursework in the biological sciences. However, 23 percent of Regular/Emeritus members reported no coursework in the behavioral sciences but only 9 percent of Associate members did. Thus the proportion of scientists trained "purely" in behavioral sciences stayed constant, but the number of those trained "purely" in biology had dropped sharply.[17]

These data are reinforced by the BGA's 1996 listing of members by their research interests.[18] Of the 577 members, 282 identified their disciplinary affiliations: 157 indicated psychology and 139 genetics. However, almost half identified more than one discipline. Of the 145 who only identified a single discipline, eighty listed psychology and thirty-four listed genetics as their discipline. In numeric terms, behavioral scientists were dominating the field.

Not only were psychologists flocking to the field, but also many were from the "clinical" subfield of psychology interested in human psycho-

TABLE 4.1. **Coursework training of members of the BGA**

Course area	Regular and Emeritus Members (N = 151)			Associate Members (N = 39)		
	Mean	Median	Standard deviation	Mean	Median	Standard deviation
"Biological sciences"	7.8	5.5	7.2	6.4	4.5	6.8
"Behavioral sciences"	9.5	7.8	9.8	10.3	10.0	7.6

Source: Plomin (1979).

logical assessment and clinical treatment of dysfunction. One psychological behavior geneticist described the intellectual climate at the University of Texas psychology department where he was trained in the early 1980s: "Behavior genetics there was, I guess what was a little unusual about it, but also Minnesota-like,[19] was the extent to which it was integrated with the clinical psychology program as opposed to genetics per se."[20] Clinical psychologists focused on the kinds of traits—personality, intelligence, psychopathology—that were concerned with the behavioral whole person. Behavior genetics was anything but popular within psychology during this period given its association with genetic claims about race differences and the limits of education for children. But what it offered to psychologists who were willing to accept this infamy was an association with "hard science" and "biology." Human behavior geneticists could marshal some of the authority of biological science without having to retreat into the lab or abandon the broad "clinical" orientation.[21]

The shift toward clinical psychology constituted disciplinary realignment not from biology to psychology but within psychology. The speaker described the ascendance of clinical psychology within behavior genetics to the detriment of the more biological, comparative (between species), ethological, and social approaches within psychology. These had been particularly strong in the field's early years and, importantly, among animal behavior geneticists. Jerry Hirsch, John Paul Scott, John Fuller, Gardner Lindzey, and William Robert Thompson were among the eminent animal behavior geneticists who worked to found the field; they were also trained mainly as psychologists and worked in psychology departments for their careers. The shift to clinical psychology (and related subfields such as education, personality, and development) was a shift to a more human-centered form of psychology.

Human behavior geneticists became empowered within the field and animal researchers weakened and isolated through these entangled disciplinary realignments. For many, the association of the "island" of behavior genetics with psychology was so complete that they came to see the field as, at heart, a subfield of psychology. One psychological behavior geneticist told me, "I guess for me the context that behavior genetics has always found itself in, for me, is as a part of psychology. And what that means, especially, you know this isn't perceptual psychology about how the eyes work, it's high level, complex-behavior psychology about why some people are depressed [and so forth]."[22] For many others

I spoke with, conceiving behavior genetics as a part of psychology was such an article of common sense that they did not state it. Animal behavior geneticists were often unhappy with this association. This can be seen sharply in one animal behavior geneticist's comment on the field's leading textbook, *Behavioral Genetics* by Robert Plomin and colleagues:[23] "It's a good book, but the title is wrong. The title is *Behavior Genetics*. It's not 'behavior genetics,' it's 'human quantitative behavior genetics.' And the animal stuff if you look in that book, is all the work that was done in the '60s and '70s . . . a lot of behavior genetics that has been done with animals in the intervening time lacks a place. It's not there."[24] The field's fragmentation and the fact that geneticists and neuroscientists had turned their backs to behavior genetics had already isolated animal researchers. But the speaker's complaint was that animal research was being literally written out of the field in this period by the human psychologists who had become hegemonic.

The Field as a Bunker

Behavior geneticists spent much of the 1970s fending off attacks from all quarters. This experience had profound effects on behavior geneticists' senses of identity and solidarity and helped to redraw the field's borders as it engendered a kind of "bunker mentality." In a description of the field's critics, a leading psychological behavior geneticist clearly depicts the behavior geneticists' sense of embattlement:

> They've got an unfair advantage. They don't do research. It's one of our main gripes against the antigenetics [people]. . . . Do research to show how important the environment is, great; do research showing that genetics is no good. But, just to attack and attack. And they can attack faster than you can ever respond. So, you say, "But okay; you're saying this is important. We've studied that. We're going to study it. It isn't." But they're so far beyond it; they don't care. I mean, they don't say, "Oh right, okay." They say, "But here are twenty other things." And you know you could never catch up because they're not doing any research. . . . It's a very negative thing, and I find that if you respond to that it's never-ending. You'll never win anyway. So, I'd rather just do the research—take a longer view of it. So, it's for that same reason I don't answer—I wouldn't answer on either side of it. But, it does, you know, sometimes seem kind of cowardly.[25]

The antagonistic relationship between behavior geneticists and their critics persisted long after the IQ controversy—the speaker is describing how the situation from the 1970s continued into the 2000s. Behavior geneticists' attitudes toward critics were linked to their perceptions that the critics were unrelentingly negative and unfair and thus unworthy of attention.

However, the antagonism to critics had another, subtler source that had to do with the emergent boundaries of behavior genetics and patterns of participation within it. It is noteworthy that the speaker depicted critics as "cowardly" and "not doing any research." More than simply being negative, on this view critics attempted to define behavior genetics from outside the field without submitting their own work to the scrutiny of behavior geneticists. Put differently, behavior geneticists viewed critics as seeking to damage scientific reputations within the field without putting their own reputations at stake.[26] For their part critics occasionally did "do research" that would fit in a behavior genetics mode, but, as I discussed in the previous chapter, much of their critique was that human behavior genetics could not be validly performed, so critique through participation did not really make sense to them. This lack of reciprocity between behavior geneticists and critics represents a breakdown of the field from the point of view of Bourdieu's theory, which holds that the essence of a scientific field is that the producers and consumers (or critics) of science are the same people, and they submit their claims to each other for inspection and recognition.[27]

Animal behavior geneticists shared something of the bunker mentality with its antagonistic relationship to those outside the field, despite not having been attacked themselves and being sympathetic with much of the critique of the field's orthodoxy. For example, one might have expected internal and external critics to have worked together based on their shared aversion to Jensen's arguments, especially the race differences claims and the interpretation of heritability scores. In fact, the links between critics were always very weak. Jerry Hirsch, the behavior geneticist most vocally critical of race research, told me that he had felt disrespected by some of the external critics. Several other critical animal behavior geneticists I interviewed expressed no love for the external critics. Although these animal behavior geneticists had criticized human researchers' work, they also voiced sympathy for the "poor guys" among their colleagues who "had been attacked very viciously at times by people who would not accept that genes had any effect on behavior. If you said that [it did], you were bad for all kinds of reasons."[28] Animal and human behavior geneticists' common defense of the field's borders and shared antipathy for its critics demonstrates the effect of behavior genetics' "fieldness," even as it was fracturing.

Many behavior geneticists used their sense of embattlement to cultivate a positive sense of identity. They were fond of telling war stories about controversies to air their serious complaints about being suppressed, threatened, and the failure of their colleagues and institutions to defend their academic rights, and also to dramatize their own strength and bravery. For example, at the fifteenth meeting of the BGA in 1986, president Sandra Scarr spoke about the field's progress and success. She told several stories about the unpopularity of behavior genetics ideas, and then gave a gripping account of a near brawl with radical student protesters when Arthur Jensen had visited the University of Minnesota where she taught. She concluded, "No action was ever taken against them [the protesting students] for abridging freedom of speech. Those were heady days, not unlike fighting the Nazis of an earlier era. Like those freedom fighters, however, we also did not win in the short run."[29] Scarr turned the old charge of fascist eugenics on its head by comparing the opponents who would suppress behavior genetics to Nazis.

The field's bunker-like quality extended beyond these subjective dimensions and became an aspect of its social structure as well. The pattern is exemplified in table 4.2, which shows the institutional affiliations of members of the Behavior Genetics Association in 1985 and 1996.[30] Behavior geneticists became highly concentrated in a small number of institutions. In 1985, over a quarter of the US members who reported an institutional affiliation were concentrated in five universities, and almost 40 percent were in the top ten. In 1996, just after the period I am considering in this chapter, the concentration was even higher. About one-third of members were in five institutions, and about 40 percent were in the top ten. There was some shift in the institutions over the eleven-year interval between lists: four slipped out of the top ten by 1996, though two of these (Texas and Indiana) were still in the top fifteen. There is reason to think that in another way the concentration of researchers at just a few institutions is even greater than these data suggest. Many prominent behavior geneticists have worked at more than one of these locations at some point in their career; there is a bit of a circuit through which many have traveled.

TABLE 4.2. **Ten institutions with the most BGA members**

Rank	BGA 1985		Cum	Rank	BGA 1996		Cum
1	Colorado, Boulder (plus, Colorado Med Center in Denver)	34	34	1	Colorado, Boulder (plus, Colorado Med Center in Denver)	50	50
2	Minnesota	14	48	2	Pennsylvania State	26	76
3	Connecticut	11	59	3	Minnesota	19	95
4	Texas, Austin	10	69	4	Washington University	16	111
5	Washington University (+ related in St Louis)	9	78	5	Virginia Commonwealth Medical College	11	122
6	Pennsylvania State	7	85	6	Veterans Admin Medical Center, Portland	8	130
7	Wesleyan	6	91	7	Virginia	7	137
8	Virginia Commonwealth Medical College	5	96	7	Case Western	7	144
8	Indiana	5	101	7	Yale	7	151
8	SUNY Binghamton	5	106	7	Indiana	7	158
	Total members	388			Total members	577	
	US	306			US	415	
	International (21% of total)	82			International (28% of total)	162	
	US no listed affiliation	32			US no listed affiliation	45	

BGA 1985
34 at 1 institution (12% of all members listing an institutional affiliation; 11% of all US members)
78 at 5 institutions (28% of affiliated; 25% of US)
106 at 10 institutions (39% of affiliated; 35% of US)

BGA 1996
50 at 1 institution (13.5% of affiliated; 12% of US)
122 at 5 institutions (33% of affiliated; 29% of US)
158 at 10 institutions (42% of affiliated; 38% of US)

Note: The counts here combine members from universities and affiliated medical centers. Some members didn't have an affiliation listed, only a home address, so the institutional concentration percentages at the bottom exclude the unaffiliated from the denominator.

In an ordinary disciplinary field we would not expect to see this kind of concentration. The distribution of researchers would be much more even since most universities would have departments. We might expect a similar concentration pattern in a field like high-energy physics where researchers are concentrated around expensive instruments. But there is no analogous situation in behavior genetics where animal labs and registries of twins and adoptees are relatively easy to set up. Although it is conjecture to say so, the field's concentration may have been an adaptation to the climate of controversy. Even after the IQ and race controversy had cooled by the late 1970s, skirmishes about behavior genetics were ongoing. When I asked about the seeming calm after the IQ controversy, one psychological behavior geneticist responded, "I don't know that I really agree that there was kind of a latent period in there [relatively free from controversy]. I mean there were a lot of other things getting worked out and established then. . . . An Irv Gottesman or a Sandra Scarr or somebody like that who was around in that period probably wouldn't see it that way [as peaceful]."[31] Another described the BGA as playing an important role as a "support group" at the time.[32] Perhaps hostility to behavior genetics made it less likely for behavior geneticists to land in the "average" behavioral science or genetics department and more likely to go where they were already accepted and where opportunities for collaboration abounded.

The idea that behavior genetics became like a bunker is consistent with its archipelagic structure and development. The bunkering effect was most pronounced among the psychologists and psychiatrists making up the islands of human behavior genetics; animal behavior geneticists were usually inclined to respect the rationale for hunkering down, though as we'll see, they bridled at some of its practical implications. The bunker effect thus helped bind together those who claimed the "behavior genetics" label. But it also helped drive further away those on islands that had already been drifting from the core of the field. In this way the bunker in one part of the field helped propel the archipelagic fragmentation generally.

Whether or not this pattern would have occurred without the climate of controversy, the concentration of behavior geneticists had several implications. One was that behavior geneticists have tended to know each other quite well and to have strong personal ties to each other. In my interviews, several of them characterized it as a "clubbish field."[33] They meant this disparagingly, and I will return to it below because animal and human behavior geneticists tend to favor two dramatically different strategies for dealing with it. Another implication is that in "holing up" in the bunker of the field and becoming institutionally concentrated, behavior genetics was coming to embody a very different form than the founders' vision of an open, inclusive field whose membership would necessarily be widely distributed. Finally, this affected the organization of scientific criticism, recognition, and trust in the field; to these issues we now turn.

Becoming Closed to Criticism

The solidarity engendered by the cultural and social "bunkerization" of behavior genetics helped preserve behavior genetics as a scientific community when it was under attack, but it also muddled norms and practices of mutual criticism and recognition among behavior geneticists. Behavior geneticists, especially human researchers, developed a persistent collective "allergy" that made them highly sensitive to criticism. It became difficult for behavior geneticists to distinguish constructive criticism from destructive attacks, and this made them less willing to engage each other critically. This allergy exacerbated tensions between animal and human behavior geneticists, since animal researchers, less subject to attack, were less allergic to criticism and more inclined to perceive practices of limited criticism in the field as problematic.

In the years following the IQ and race controversy, behavior geneticists progressively disengaged with critics from outside the field. In the mid-1970s, *Behavior Genetics* published several critics' letters targeting articles in the journal and responses from the authors.[34] But by 1978 such exchanges stopped appearing. Behavior geneticists came to ignore critics' contributions, as a psychiatric geneticist explained:

> And so the concerns that [population geneticist and critic] Marc Feldman expressed way back about the nature of the heritability statistic and the fact that it's a local parameter that may only be true when there's linearity, or it's only going to be true under very restricted conditions, was very foresighted. And it's just taken people in behavior genetics and psychiatric genetics a while to appreciate what seemed like rather arcane objections that many people thought were partially motivated by, you know, special interests or a particular sensitivity to racial issues or issues about intelligence. But in fact they turn out not to be idiosyncratic or politically oversensitive, but to really be at the heart of what you have to face with you deal with complex phenotypes.[35]

This statement is especially noteworthy because it comes from a long-standing champion of the field. The speaker sees the field as having arrived, three decades later, at a position that behavior geneticists had long dismissed as a critic's special pleading.[36] In 1986, behavior genetics' most famous critic, Richard Lewontin, coauthored the book *Education and Class: The Irrelevance of Genetic Studies* that showed there was no genetic determination of the IQs of a set of French adoptees when social class was controlled.[37] Although this work would certainly meet any substantive definition of a behavior genetics study, it failed to penetrate the literature.[38] Indeed, the complaint cited earlier that critics do not make positive contributions and "don't do research" was as much a matter of behavior geneticists' selective perceptions as critics' actions.

Just as behavior geneticists progressively ignored external critics, they marginalized the few inside the field who dared to take strong critical stands. This is what happened to Jerry Hirsch. Despite his centrality in the field's early years, behavior geneticists came to ignore him as his critiques became increasingly strident of Jensen, his allies, and the field generally for adopting aspects of Jensen's agenda.[39] Several in Hirsch's circle ceased active participation in the field.[40] Hirsch became perceived as "a little funny"—overly emotional, perhaps slightly unhinged—and eventually many behavior geneticists would no longer recognize his work as part of behavior genetics, though it had not changed substantively.[41]

The implicit injunctions against internal criticism extended well beyond vocal figures like Hirsch. One animal behavior geneticist told me that as behavior geneticists circled the wagons against outside critics, they also refrained from taking critical positions with each other. Criticizing each other, he says,

> was completely not done . . . so the discussions at those meetings [at the BGA for example], there was never a critical question, never really critical. . . . There was kind of this mindset: don't criticize each other. And, in that sense, that was clubbism . . . you stand by each other, and you don't hang your dirty laundry outside for people to see.[42]

The speaker described running up against this norm several times when he tried to get comments published on articles that had appeared in *Behavior Genetics*. Once as a somewhat over-eager graduate student (circa 1983), he wrote a letter pointing out some conceptual and mathematical errors in an article. The letter was rejected, and he interpreted the editors' rationale like this: "These guys [the editors and reviewers of his letter] were saying something like, this is an honest effort to analyze behavior and basically you shouldn't criticize that. . . . That was kind of the idea I got. I was criticizing somebody within the group and you don't do that."[43] In another letter criticizing an article's effort to extend the concept of heritability, the editors responded that the points he raised were valid, but would be obvious to any reader of the journal. He wondered, "Okay, so why did the original article get published?"[44] Their rejection concluded, "This is not a journal dedicated to criticizing methods and statements of individuals making serious efforts to understand behavior."[45] The editors' response indicates that within the tight behavior genetics network, members felt they could trust each other's judgment on such matters. An unknown graduate student, like this speaker at the time, pointing out mathematical errors and imprecise language and questioning assumptions would immediately appear to be engaging in the kind of destructive nitpicking seen as characteristic of the field's enemies.[46]

A psychological behavior geneticist who researches human intelligence described a different version of the problem that emerged when he was at odds with colleagues who had supported *The Bell Curve*:[47]

> I wasn't sure I was going to get tenure, I wasn't getting a lot of work out, I didn't agree with the so-called mainstream view of things. . . . [But] I started getting the work out that allowed me to have a career path as someone who accepted some of the assumptions of the field but didn't accept a lot of the conclusions. . . . It wasn't clear at the time that that was going to happen. So it was a time that I felt disillusioned.[48]

Unlike the previous speakers, this researcher was not complaining about the difficulty of getting an article published that criticizes the work of other behavior geneticists. Rather, his problem was that the intellectual space would not permit alternative forms of research that might be perceived as "critical." He wanted to do behavior genetics that criticized its assumptions, without being rejected as another Leon Kamin. Further, it wasn't that there were things everyone knew but simply couldn't say; rather, it became almost impossible to imagine an alternative or to interpret sharp questions and deep challenges as being put forward in a constructive spirit.

It is important to realize that through all this, behavior geneticists participated in all the ordinary institutions of peer review and open communication. There was no conspiracy to suppress critical voices, nor were behavior geneticists acting in some patently unscientific fashion.[49] In their tightly knit network, behavior geneticists believed they understood the scientific limitations (of heritability, of twin and family studies, of generalizing between human and animal behaviors, of defining traits) and had nothing to gain from airing these issues again. As one of the frustrated critics described the situation:

> [The critics' presentation] allows them [behavior geneticists] to dismiss all this criticism as the rantings of some extremists. Because they *are* ranting. [laughs] That looks very reasonable. But the substance of what they're saying is getting lost. . . . I didn't have the feeling at that point that there was any communication going on any more. It's completely ritualized, restating of old positions. . . . To get back to the race thing. People don't want to think about that. Because if you think about it then you have to say something, you have to have an opinion. It's much easier to not think about it, not have an opinion, and you don't get yourself into trouble.[50]

The speaker was describing a situation where it had become nearly impossible to disentangle the novel critique from the known, the constructive from the destructive, the rational from the emotional, and the scientific from the political. With these ambiguities behavior geneticists' allergy to criticism was hardly unreasonable.

A crucial dimension of the ambiguities of mutual criticism was the conflict between human and animal behavior geneticists. The story of frustrated internal criticism is largely about animal researchers commenting on the work of human researchers, rarely human researchers criticizing each other, but never human researchers scrutinizing animal research. For the animal behavior geneticists there were two main problems with the situation. First, the issue wasn't so much the substance of the critique—whether or not this explanation of heritability was sloppy or that the estimates of genetic links to aggression had been inflated, for example—rather, it was that this implicit censorship indicated something dysfunctional about the field's culture and practices of communication. Second, they resented the fact that criticism implicitly located the critic at the margins or outside the field, as one of its enemies rather than one of its friends. Indeed, critiques tended not to appear in journals that published behavior genetics regularly but in those outside.[51]

An important effect of this conflict has been to spur ongoing mistrust between animal and human behavior geneticists. A mouse researcher voiced his suspicions to me:

> And my gut feeling is that most human behavior geneticists have no idea about genetics. They don't know Mendelian genetics. They have all these

fancy models but there is a gap in the science. I can't prove it; I have my own things to do. I don't want to spend my career proving that someone else is wrong. . . . [When following their literature more closely] I kept feeling like it itched, it itched.[52]

Another animal behavior geneticist proclaimed her opinion more boldly when describing a textbook she was working on:

There's going to be less human stuff in this book. Screw it, you know what I mean? When the tests are in question, the statistics are in question, the populations are in question, and the interpretations are in question—I hope that's on your tape—I mean what's left? I remember once when I was younger, much younger than you, I swam for an Olympic judge, and he told me there was nothing wrong with my swimming except my kicking, breathing, and stroking. [laughs][53]

As these remarks display, animal behavior geneticists began to harbor epistemological doubts about their human researcher colleagues that were fueled, in part, by the norms against some forms of criticism. The mistrust was not symmetrical, however. Human researchers did not have any particular epistemic misgivings about animal researchers; indeed they did not ordinarily pay them much attention.

One might argue, following Karl Popper and Robert Merton, that unbridled "conjectures and refutations" and "organized skepticism" are the very essence of science.[54] However, more recently, sociologists of science have noted that doubt is infinite in principle, and science can only proceed through usually implicit norms and practices curtailing critique, what Bourdieu would call the legitimate "censorship" imposed by the dominant.[55] The problem for behavior genetics was that the situation of controversy made the proper balance impossible to find, so the censorship that in most fields is implicit and legitimate became sometimes explicit and contested and thus corrosive to trust.

Conflicting Visions of the Field

These developments—the disciplinary realignment of the field and its bunkerization—were the outcome of behavior geneticists' collective adaptations to the IQ and race controversy and the two decades of somewhat less intense conflict that followed. All behavior geneticists were affected, but human and animal researchers were affected differently. In response they adopted different visions of the field that have carried on to the present.

In many ways, human researchers benefited from these developments because the field came increasingly under their control, its arrangements helped them reproduce themselves and protect their scientific authority, and it linked to disciplinary interests in behavioral science. But the safe space they had built began to feel constraining to some. One psychological behavior geneticist, a leader of the field, described the situation this way:

> [At the BGA meetings] we all agree with each other and we know where they're sort of going. There are some new things that come up, but, you know, I find out about it anyway. So I don't really need to go to those meetings. But, I think the general point I'm making is an important one for the field. . . . I always have felt intuitively that we need to have a support group, especially during those hard times. But now that it's not particularly hard times, I think we can do better by giving the field away and getting other people to do it.
>
> You know I've published a few things in *Behavior Genetics*, but no, it's always been my goal to be a developmental psychologist who does genetics. It's my, probably my message in terms of giving away the field. I don't want it to be a specialty—well it can be a specialty field for people who have the methodological skills, but I want to give it away in the sense that I try to tell people, "You don't have to be a geneticist to do this stuff."[56]

As this speaker explained, his key interest has been using behavior genetics to do psychology better. He noted the close-knit character of the field and acknowledged its function as a "support group," but he came to see this as limiting behavior geneticists' possibilities. Behavior geneticists were familiar with each other and "all agree," and this made him unenthusiastic to engage with them. He would prefer behavior geneticists to conceive themselves less as members of a field and more as bearers of tools that can be shared with researchers in other fields.

A leading psychiatric geneticist who has participated in behavior genetics since the mid-1970s explained his "coolness" toward behavior genetics this way:

I don't really actively think of myself as belonging to any club. I'm really just trying to work on the different problems, and—and yet you know I'll attend meetings and I have friends, but I try not to maintain a union card because I think that's really dangerous. I try to grow with the problems as I see it, and if you get too caught up in the assumptions and the traditions of any one club, then it can bind you. . . . What happens is that people get—they acquire a skill and then they tend to like to just continue to do what they're good at.[57]

On the one hand, as this speaker explained, because he is a psychiatrist not a psychologist, he did not feel fully "accepted" in behavior genetics. But on the other hand, he claimed he would not want to be too intimately associated with behavior genetics or any other "club" because they are all bound by collective assumptions, traditions, and methods. Instead, he tries to align himself with "different problems"—for example, schizophrenia or personality disorders. While he did not target behavior genetics as an especially problematic club to be a part of, the longer narrative of his interview suggests that much of his perspective on the problem of clubbishness arose during his long engagement in behavior genetics.

These speakers see participation in behavior genetics as a *means toward other scientific ends.* They see too deep an investment in behavior genetics' conventions and tools as intellectually constraining. And they express a sense that the scientific recognition of behavior geneticists can be taken for granted ("we all agree") or is at least predictable and not overwhelmingly desirable (as when the psychiatrist discusses his reception). But their ambivalence should not be seen as a rejection of behavior genetics, since both these speakers have invested tremendous time and energy as participants in the field and advocates for the science. Rather, they prefer to see engaging behavior genetics as a means toward other ends, specifically, competing for scientific rewards, or capital, in other fields such as developmental psychology and psychiatry and engaging particular biobehavioral problems like psychopathology and the origins and development of personality and intelligence. This orientation to the field is typical of human researchers, and it is ironic that those who dominate the field are ambivalent in their commitment to it.

Many animal researchers also became dissatisfied with the field's development during this period, but rather than counseling detachment, they have advocated collective reinvestment and reinvigoration of be-

havior genetics' social and epistemological institutions. One interviewee
bemoaned the fragmentation of the field:

> I don't want IBANGS [International Behavioural and Neural Genetics So-
> ciety, another behavior genetics society] to become the mouse meetings.
> There are psychiatric genetics meetings, [the] BGA and it's 90 percent human
> quantitative genetics. I would love to have them come to our meetings too.
> I would love to have all those meetings together. At one point I tried that. I
> proposed a joint meeting of the ISPG, the International [Society of] Psychi-
> atric Genetics, BGA, and IBANGS. And everybody said "it's a nice idea,"
> but then nothing happens, nobody gets back to you. Because they're not re-
> ally interested.[58]

This speaker echoed those above in criticizing the clubbishness of behavior genetics and the tendency for communities to coalesce around small
distinctions, which formal societies tend to exaggerate. What's different
is his proposed solution: he sought to bring the various societies together,
to foster exchanges and communication by holding all the meetings simultaneously. Another animal researcher focused on the epistemological problems he perceived with the field's fragmentation, clubbishness,
and wariness of criticism:

> It would be nice if all of the people in the different areas who want to study
> these things can kind of get their act together about some common frame-
> work for posing meaningful questions. And that would be my kind of dream
> of dreams if a lot of people could agree on, "look these are just not legitimate
> methods," if we could have a laundry list of things that we all agree are basi-
> cally not going to tell us very much . . . [and if] we could also kind of agree on
> the things that we think do tell us something . . . then I think we could have a
> very good basis for creating new ways of asking the question. But that would
> be my minimal hope, not that these kinds of distinctions [between positions]
> would go away but that we could be a little better about finding ways of put-
> ting our energy into meaningful venues for getting answers to them.[59]

He envisions a kind of summit meeting in which behavior geneticists
and fellow travelers would break through the disciplinary differences,
the tendency to become closed to outsiders, and the distaste for criticism
and hash out a common framework that would differentiate illegitimate
from meaningful science. We might be skeptical of these speakers' diag-

noses of the problems and their proposed solutions—that lack of interest has kept people apart and that getting everyone in one room would lead to agreement (implicitly with the speaker's scientific view). But the visions they articulated reflect a very different orientation toward behavior genetics. They see it as an *end in itself*. Animal behavior geneticists have been much more concerned about developing behavior genetics as a coherent set of scientific practices and ideas and as a distinctive, mutually interested research community—that is, hoping to recover the vision of the 1960s—not merely as a portable set of research tools or an accidental set of scholars happening to share interests.

Animal researchers' orientation toward the field as an end in itself is also reflected in the motivations they express for doing research. Whereas human researchers tended to emphasize the solution of pressing social problems or other fields' intellectual puzzles, animal researchers more often emphasize intrinsic interest. One speaker explained his motives this way: "And as a person who does mostly animal-based research, I now do some human research, I can honestly say what drives me has been more an interest in how things actually work. The idea that people can be helped by that is great, but that's pretty much my fundamental interest."[60] When I asked another animal researcher why some in the field are interested in studying racial differences, she replied:

> It's to get at the basis for certain human mental traits. That's what you're interested in. . . . That doesn't mean the subject is useless. To me it's more interesting to study the genetic basis for taking a PhD in sociology. Is there something genetic similar about all doctoral students in your subject? It's just interesting for its own sake. I don't care about the color of your skin; it's totally irrelevant.[61]

These speakers assert that they are motivated by understanding the phenomenon, not what that understanding might enable. The example of the genetic influences on interest in sociology is humorous, but the speaker was asserting that the social irrelevance of the trait is part of what makes it scientifically interesting. By implication, it is the intense social and political interest in racial differences that clouds the science and makes it uninteresting. Thus being oriented toward behavior genetics as an end in itself is both about an approach to scientific topics and the field's community.

These two visions of the field correspond also to different under-

standings of behavior genetics as a science. A human psychological researcher explained this difference. He expressed sympathy with the frus-
trations of some animal researchers but contrasted their views of science
with his own:

> I'm saying, we're still doing old-fashioned social science with a little bit of ge-
> netic information added in, and there are a lot of limitations to doing that and
> let's just keep doing the best we can. They [some animal researchers], I think,
> concluded that you can't do real science on these things [human behaviors],
> genetics or no genetics, and you shouldn't even try. And therefore they do
> low-level mouse work. And they've kind of said rather than being commit-
> ted to the phenomenon, the complex human phenomenon the way I am . . . it
> always seemed to me that they're committed to the scientific methods. They
> want to be real scientists, so they need to find a phenomenon for which they
> can do something that seems like real science. And that is lower-level experi-
> mental animal work for the most part.[62]

This speaker is invoking what we might call, inspired by the philosopher Hans-Georg Gadamer, a notion of "epistemic gain."[63] Gadamer argued that objectivity is impossible and knowledge is always embedded in
a cultural horizon, thus epistemic gain—a criterion of intellectual progress without "truth" as an ultimate goal—is the point of knowledge. This
speaker is saying that genetics offers epistemic gain to the understanding of human behavior, but he believes that animal researchers, invoking an absolutist understanding of science, reject human research as too
messy to yield real truths. Although the speaker perhaps overstates animal researchers' rejection of the possibility of valuable human research,
the different visions of science are clear. The animal behavior geneticists quoted above were concerned with creating a canon of methods to
distinguish "legitimate" from "illegitimate" techniques. Animal behavior geneticists' tend to want to police sharp division between meaningful
and meaningless science; human researchers express more tolerance for
incremental improvements in messy problems.

The practical implications of these two visions of the field were different for human and animal behavior geneticists. For human researchers
it meant behavior genetics was one of several scientific attachments, and

it usually took a secondary role to a disciplinary identity. Participation in behavior genetics was a means for them to compete for scientific recognition in other fields, while they have been able to take each other's recognition more or less for granted. These competing interests tended to limit their investments in behavior genetics (of time, energy, identity, scientific reputation). And the collective disinvestment limited the rewards for investing in the field, which, in turn, reinforced people's tendency not to invest deeply. The multiple competing interests, without the field being a strong center of gravity, have also encouraged the narcissism of small differences among members—which is to say they are often at pains to distinguish themselves from others whom their methodological tools and substantive interests overlap considerably.

Animal researchers have adopted a range of practical adaptations. Some have completely disengaged from the field because of the shortcomings they perceive. For example, one interviewee, an animal researcher who was a graduate student during the late 1970s, described his feelings after the IQ controversy:

> I got very disappointed and sort of demotivated at the way that everything was just immediately polarized into politics. And when I left graduate school I left it [behavior genetics], I said, "I cannot do this anymore, it's a waste of time, its not getting anywhere." . . . As soon as it got polarized, the whole issue became a nonissue to science. And other people working in other fields went on about their business and advanced our genetic understanding of how the brain works and there you go.[64]

Individuals like this one are among the residents of those islands that "ain't called behavior genetics." Others, like the speakers above, have remained in the field, sometimes marginalized, and have tried to work on their research while trying to effect changes. And still others have remained in the field uncomplainingly. But why would these animal behavior geneticists remain committed to the field (whether articulating critiques of its deficits or not) if they are unable to realize their preference for the robust field and if their association with behavior genetics hinders their capacities to seek recognition from outside the field, as do their human researching colleagues?

For many, staying in the field was not exactly a "decision." As Bourdieu reminds us, professional identity is part of the durable dispositions that make up the habitus.[65] Behavior genetics was the scientific game in which they'd invested and knew how to play; switching fields, even if it might seem to be the "rational" course of action, is not as easy as changing your clothes when they become unfashionable. A few animal behavior geneticists chose to adapt by taking on human research in addition to animal research. In the 1996 survey of BGA members' interests, 161 members (of the 290 who reported their interests) studied only humans, eighty-three studied only nonhuman animals, and thirty-five studied both.[66] Of this thirty-five, I am familiar with the careers of ten; nine of them were primarily animal researchers who later added human research, while only one moved in the other direction.

Furthermore, animal behavior geneticists were also able to maintain a reasonable degree of control over the offices and honors of the Behavior Genetics Association. Although human researchers outnumbered animal researchers by about two to one in 1996, sixteen of the thirty-eight BGA president positions and fourteen of thirty-three Dobzhansky Awards (the BGA's distinguished career award) have been held by members who've done significant animal research.[67] The society has maintained a rough pattern of alternating, though in the last ten or fifteen years this has skewed to favor human researchers more heavily. Animal research has also always featured prominently in *Behavior Genetics*, the society's journal. What this means is that animal behavior geneticists' investment in the field has paid off, to a degree. However, this recognition has been most reliably that of other behavior geneticists rather than that of outside audiences, and this contrasts with the human behavior geneticists. A clear sign of this is the Institute for Scientific Information's list of "Highly Cited" researchers. In late 2006, there were at least fifteen people on the list (of several hundred) who would likely identify as behavioral or psychiatric geneticists, but all of them are human researchers.[68] There are powerful, highly cited animal researchers who study genetic influences on behavior, but they do not identify with the field of "behavior genetics."[69] Thus while some animal behavior geneticists abandoned the field and those who remained did so with differing senses of having given something up to do so, they managed to secure what is in some sense a disproportionate amount of scientific recognition and rewards within the community (even if they have felt frustrated in their capacities to set the agenda). Human researchers, in contrast, have leveraged their association with the field for rewards in other fields, with some, at least, being very successful.

Overall, this situation is highly ironic from the point of view of Bourdieu's theory. In most fields the dominant members are autonomously oriented. Reaping the field's greatest rewards they tend to be most invested in its integrity and vitality, while weaker members tend toward heteronomy, pursuing recognition in other fields. In behavior genetics the pattern is reversed. The subordinate animal researchers want to strengthen the field and thus their authority while human researchers instrumentalize the field as a platform for their pursuits elsewhere. The next chapter considers this story in greater depth.

Narrowing the Field's Intellectual Possibilities

The founders of behavior genetics had envisioned the field in universal terms. They fought strong associations with particular disciplines or paradigms and hoped that all scientists whose research fit the substantive goal of linking behavior and genetics would find a place for their research. But as behavior genetics became closed, fractured, and de facto disciplinary ties strengthened, the range of scientific topics it contained became narrower. I consider in a moment how behavior geneticists became averse to certain kinds of theorization. But first I show how behavior genetics became focused on the analysis of *individual differences* or the explanation of sources of variation in behavioral traits. This came at the cost of several other possible research directions that had been or could have been part of it.

One possible, but avoided, pursuit was the question of racial comparisons in behavior. Despite the flurry of action during the IQ controversy and the endorsement of the genetic hypothesis for explanation of the black/white IQ gap, the mainstream of behavior genetics pushed this issue to the field's margins. There remained an undercurrent of interest among a small number of researchers (Arthur Jensen, Philippe Rushton, Richard Herrnstein, for example), but most would echo what a leading psychological behavior geneticist told me, with a note of exasperation, "It's just such a small part of behavioral genetics, and I really don't think that there are tools [for reaching solid conclusions]."[70]

Another topic that mostly disappeared, or never got going, was evolutionary analysis and theorization about behavior. In the field's early years, behavior geneticists thought evolution would offer a unifying theoretical framework and an opportunity to make sense of comparisons among species and among cultures.[71] Behavior geneticists would occasionally draw on evolutionary argument to justify their work.[72] Sandra Scarr, in her 1986 presidential address to the BGA, said, "we are no longer an embattled minority fighting to keep the light of evolutionary truth burning."[73] But despite this rhetoric, evolutionary theory did not become important in their research.[74]

One reason for the dissociation was the controversy sparked by the publication of E. O. Wilson's *Sociobiology* in 1975 and then Richard Dawkins's *The Selfish Gene* in 1976.[75] Wilson's book was about the evolution of behavioral traits in species from ants to apes, but in a final, admittedly speculative, chapter he argued that human behaviors—in particular traits like xenophobia, war, and differentiated sex roles (for example, male aggression and female nurturance)—were the product of adaptive evolution. Dawkins's argument that all of life should be interpreted as individual genes battling to propagate themselves was not centrally about organisms' behavior, though it did seem to naturalize aggression and traditional gender roles, but it furthered the view that environments don't much matter. These views were highly complementary with the IQ and race controversy, which they followed closely. To the public, and many critics, these ideas seemed to be generating a coherent story about the biological causes of social problems and the fundamental limits of political efforts to bring about socially equitable change. Critics like Lewontin, Kamin, Rose, and Gould decried these ideas as symptoms of the rise of a new pan-scientific revival of biological determinism and the acquiescence of biology to political ideologies from which scientists had long struggled to free themselves.[76] Rather than fighting on these new fronts in their war with critics, behavior geneticists became more assertive in defining their field as concerned with the genetic *and* environmental causes of individual *differences* in human behavior and that the analysis of evolved *universal* behaviors were not their concern.[77]

Behavior geneticists used this definition of the field as concerned with individual differences to fend off another attack from the point of view of developmental biology. During the IQ controversy, one of the critiques of the use of heritability estimates by behavior geneticists generally and Arthur Jensen in particular was that they didn't capture the interaction between genotype and environment to produce phenotypes.[78] As a defense, Jensen distinguished between genotype/environment interaction, which concerns sources of variation in a population, and "interactionism," which concerns the mechanisms that lead to individual development.[79] This point was reiterated and developed by several behavior geneticists as well.[80] The aim of the argument was to define behavior genetics so as to make it less vulnerable to a certain critique, but the effect of this was to define behavior genetics as pitched *against* biological developmentalism. In the 1960s vision of the field, however, both sides of Jensen's distinction would have been seen as part of the field.

These defenses of behavior genetics and the ways they served to refocus debates about the field and the scientific activity of its members simultaneously narrowed the purview of mainstream behavior genetics and heightened conflict between human and animal researchers. Many animal behavior geneticists had been inspired by the tradition of ethology, the field devoted to the study of animal behavior and in particular the comparative study of species to build generalizations about behavioral patterns and causes.[81] The work of Wilson and Dawkins can be seen as an effort to bring (or at least enforce through dissemination) a stronger logic of evolutionary reasoning to ethology. So in drawing boundaries against this work for reasons of public controversy management, behavior geneticists were also marginalizing ethology-oriented animal researchers. A similar dynamic happened in the cordoning off of developmentalism. Animal experiments were particularly well suited to this kind of research, so, again, marginalizing it pushed aside animal researchers in the field. It is worth noting that John DeFries was one of the architects of this boundary, and he was also an animal researcher who shifted much of his research portfolio over to human studies (particularly of reading disability). But other animal behavior geneticists found these boundaries wielded against them. When I asked one eminent human behavior geneticist about the role of Jerry Hirsch, the animal behavior geneticist and implacable critic of the field's mainstream, he replied, "Jerry Hirsch was not interested in within-species differences so much as he was interested in between-species differences. That is not behavior genetics."[82] Here the "individual differences" focus was a tool to police another member. But more routinely, these boundaries simply served to mark limits to the recognition animal researchers could hope to receive.

These boundaries weren't exclusively about animal researchers. In the field's early years, there had been human researchers interested in bringing ethological and evolutionary/comparative analysis into the field. For example, Daniel Freedman combined ethology with twin studies in his studies of smiling and fear of strangers in infants—behaviors he thought were adaptively evolved.[83] And animal behavior geneticist Benson Ginsburg and anthropologist William Laughlin called for cross-cultural comparisons of genetic connections to language acquisition and adaptations to different physical environments.[84] But these approaches came from disciplinary locations—namely, infant development psychology and anthropology—that lost their footholds in the field as the clinic-oriented psychometricians took control.

A final intellectual consequence of the post-IQ controversy era was that behavior geneticists became averse to theoretical rumination about the scientific meaning of their research. One animal researcher described a volume called *Theoretical Advances in Behavior Genetics* from a conference held in 1978:[85]

> This, as far as I'm concerned, is the last time that people thought about behavior genetics . . . there have been no theoretical advances in behavior genetics since. . . . [Pointing to the contributors:] These were human and animal people, this is a human person, these are more animal guys. They are sitting together and thinking together. . . . And that's a pity, this is what's missing in the field of behavior genetics. It was twenty-six years ago, the last time people talked about what does it mean, what are we doing? And they don't do it; it's as simple as that.[86]

A human behavior geneticist echoed his thoughts:

> there's just no niche in academic psychology for that kind of . . . perspective taking, and particularly in behavior genetics. I mean another reason for that in behavior genetics is that to this day I think people associate any kind of theoretical, philosophical work with getting slammed and getting called a racist. There's no tradition of relatively sympathetic theoretical treatments.[87]

Several of the quotes from earlier in this chapter also pointed to deficits in theorization or perspective taking in behavior genetics—for example, explicitly in the one about the "lack of venues" for discussing "meaningful" approaches and implicitly in the quotes about ways the "club" was limiting. It is noteworthy that this perceived lack was not only the perspective of animal researchers and those who would like to cultivate the field as an end in itself. Also they saw the theory deficit as a general problem in their region of the behavioral and biological sciences but particularly acute in behavior genetics. These speakers are pointing toward a prevailing form of intellectual anomie—the lack of norms and opportunities for abstract thinking about the science.

One consequence of the disregard for theory was that behavior geneticists haven't pursued some anomalies deeply. For example, behavior geneticists often hold up the Texas Adoption Project as a strong example of an adoption study. It shows that the IQs of adopted children are well correlated with their birth parents but poorly with their adoptive parents, which suggests a strong genetic relationship and yields heritability estimates of 0.45 to 0.53.[88] But philosopher Jonathan Kaplan has pointed out that in the data, adoptive parents' IQs were no more correlated with their own children (with whom they shared genes and environment) and that IQs of mothers who gave up children were significant correlated with their children's adoptive siblings (though they share neither genes nor environment) and sometimes to a greater degree than adoptive parents had to their own children. Kaplan writes, "such bizarre results undermine the simplistic interpretations that are given to those results that are more easily explained."[89] And apparently behavior geneticists have not taken up their meaning.

Another anomalous example that hasn't been explored is the relationship between the heritabilities estimated in animal studies and those in human studies. One animal behavior geneticist told me that heritabilities for behaviors reported in animal studies are almost never as high as those reported in human studies.[90] But this is confusing; one would expect the reverse since animals raised in strictly controlled conditions should experience less environmental heterogeneity than humans who grow up in the world, thus the environmental portion of trait variability should be lower and genetic sources higher. This has been pointed out from time to time. Based on his studies of dogs' learning, Scott wrote, "Compared with similar heritability figures in human research, especially those for intelligence tests, the human figures appear to be far too high."[91] Scott also said that dog breed differences in behavior were partly due to puppies being raised in litters of similar dogs and that this might help explain some of the characteristics of human twins. Whether or not Scott is right, his ideas signal possibilities for putting human and animal research in closer dialogue that, to my knowledge, behavior geneticists have not pursued.

A third example concerns puzzling difficulties animal researchers have sometimes had trying to replicate each other's work. In 1998, John

Crabbe, Douglas Wahlsten, and Bruce Dudek tried to repeat a set of experiments in their labs, respectively, in Portland (Oregon), Edmonton (Canada), and Albany by precisely controlling the strain of inbred mice, their age, time of day, the cages and food, and handling and testing protocols for several common behavioral tests.[92] Despite heroic efforts to control differences, they found test results varied widely between sites—a finding that militates strongly against the genetic determination of behavior.[93] The remarkable sensitivity to conditions is a problem that animal researchers are now beginning to deal with. But do these experiments cast any light on human research? Are humans as sensitive as mice? Does this experiment help explain why it has been difficult to get consistent associations between particular genes and human behaviors (see chapter 6)? Does it have implications for the claim, common among human behavior geneticists since at least the IQ controversy, that apart from fixing serious deprivation, ordinary environmental interventions aren't likely to affect educational gaps or improve social problems? Conditions in behavior genetics make it difficult to pose such questions.

At some level behavior geneticists have expressed concern with the field's undertheorization for its entire history. Recall, for example, Daniel Freedman's concern that behavior genetics had a "gimmick" rather than a theory and his call for evolutionary theory to plug that gap (chapter 1). But the shortcoming intensified after the IQ controversy. Part of it was the two-sided defensiveness about speculative argument. On the one hand, most theoretical arguments behavior geneticists have experienced are critiques of their work. On the other, when behavior geneticists speculate about the meaning of their work, critics have often attacked them for overreaching. The anxieties about criticism generally have hindered theoretical discussion. Another part is the field's intellectual fragmentation and the identification of "behavior genetics" with a limited set of possibilities. This means there are few opportunities for behavior geneticists to engage deeply the theoretical implications these alternatives pose, and, more crucially, theoretical talk usually signals conflict between different intellectual communities.

The sloughing off of different intellectual perspectives and the anxieties about theory served to impoverish the possibilities for human/animal comparisons, which never became systematic or more than gestural. Humans, of course, are animals, as anyone would quickly point out. Yet the field's divisions have left this fact little more than a truism. Nothing about the research would lead one to think humans' animal nature matters. But more generally, here we can see that the field's fragmentation into a disciplinarily differentiated archipelago and the boundaries produced by these intellectual conflicts mirrored and reinforced each other. In the end, what came to count as behavior genetics was cognitively and disciplinarily much narrower than it had been before the IQ and race controversy.

Conclusion

So, should researchers use animals or humans to study the genetics of behavior? In the field of behavior genetics, the answer has become "humans." Although in the years of the field's founding, behavior geneticists imagined a dynamic and interdependent relationship between animal and human research, after the IQ controversy the two approaches drifted apart and human research came to dominate the field's agenda.

But this "answer" wasn't the result of the scientific controversy being resolved. There were no experiments, for example, to coordinate and compare the experimental systems of human and animal research. The answer was the product of the human/animal question becoming entangled in the conflicts and transitions in the aftermath of the IQ controversy. As I showed here, beginning in the mid-1970s, behavior genetics underwent a disciplinary transformation wherein the field became closely aligned with behavioral sciences, especially clinical psychology research, and dissociated with more biology-oriented disciplines. Fending off the intense attacks over the IQ controversy, behavior geneticists adopted a bunker mentality and drew a particular set of boundaries to protect the individual-differences research paradigm from attack. Even as animal researchers honored the implicit pact and participated in the defense of the field, these changes served to disempower and marginalize them.

These changes to the field provided this answer to the human/animal question, in essence by preventing the dispute from being aired in intellectual terms. The disciplinary realignment meant that many animal researchers left the field while different researchers who flooded in were from a disciplinary location and part of a network that wasn't much concerned with the question. Further, the collective defensiveness about criticism made it difficult even to pose the question without it being perceived as an attack on the legitimacy of behavior genetics. Other research communities that study links between genes and behavior but "ain't called behavior genetics" certainly answered the question differently. But in the fractured, archipelagic scientific space a genuine scientific exchange could not take place.

In this chapter I have shown how the anomie, which animates misbehaving science, became institutionalized within behavior genetics. The vision of the field initiated by its founders and championed by animal researchers of an integrated transdiscipline devoted to the genetic inheritance of behavior foundered. Ascendant was human researchers' vision of a multidiscipline where scientists from different disciplines could arm themselves with the means to do intellectual battle in their home fields. One dimension of anomie, then, was that no shared disciplinary standard would govern the scientific conduct of scientists in the field, and another was that the pursuit of recognition in other fields and the ability to take recognition within behavior genetics more or less for granted meant that there was less pressure for them to develop converging standards or challenge each other through mutual competition. Animal and human researchers' competing visions of the field also represented different solutions to the problem of ongoing controversy. The former would have sought to institutionalize vigorous internal conversation, collective criticism and policing, and mutual responsibility through deep investments. The latter did so through a bunker mentality. In effect the pact between animal and human researchers that emphasized mutual defense and responsibility became stronger yet one sided: Behavior geneticists would defend each other without critical interference. Anomie was thus directly promoted through mutual noninterference. The lack of expectations or opportunities to connect one's ideas to alternate approaches or frameworks was a more general and subtle driver of anomie. Anomie, therefore, was not total chaos, but a set of rules, structures, and barriers that institutionalized noninterference, nongovernance, noncompetition, noninteraction, and nonresponsibility as powerful forces among behavior geneticists.

None of this can tell us what would have happened under different conditions. But it is at least possible that the relationship between human and animal research could have been very different. Researchers could have been much more interested in their complementarities. The topical breadth of behavior genetics might have been much wider. A more diverse range of researchers and topics could have been included under the label, the implicit pressure to conform to intellectual norms might have been lighter, but, more importantly, there could have been alternate intellectual synergies. For example, philosophers James Tabery and Paul Griffiths have excavated the paradigmatic differences between the biometrical tradition of many human behavior geneticists and the developmentalist tradition of many of their critics to explain the intractability of the controversies about heritability.[94] But the idea that these traditions are alternative and opposed to each other rather than complementary isn't given solely by their intellectual structure. The founders of behavior genetics sought the complementarity, and at least some animal behavior geneticists—especially those frustrated by the field's form—have drawn from both sides. What is available to scientists could have been otherwise.

This chapter raises a new question: how were behavior geneticists, human researchers in particular, able to maintain themselves as scientists and continue their work? Having ensconced themselves in a protective bunker yet rejecting many of the disciplinary institutions and intellectual functions of traditional fields (which is what many animal researchers pined for), how did they build scientific authority and secure scientific resources? These questions underlie the controversy about how genetics should be incorporated into the behavioral sciences, which we explore in the next chapter.

CHAPTER FIVE

The Power of Reductionism

Valorizing Controversial Science

The nature versus nurture controversy is closely related to the debate about reductionism in the explanation of behavior. One side argues that you haven't really explained something until you reduce it to the interactions of its parts. Behavior is a product of the brain, brains are made up of interconnected neurons, and neurons develop and function through gene-guided processes. The core commitment is to building explanations from the most fundamental elements, even while acknowledging that the causes and interactions are more complex. The opposed view is that complex systems have emergent properties that are neither predictable nor explicable from their elements. Cultural norms, opportunities and incentives, interactions and situations, and learning over time all crucially shape people's behavior. Brains and bodies are necessary conditions but utterly insufficient to understand how these emergent properties affect behavior. This opposition maps roughly onto nature versus nurture: reductionistic explanations tend to be about nature, and emergent ones tend to emphasize the nurturing context.[1]

An important part of the nature versus nurture debate has always been about the relationship among the sciences. Nature advocates tend favor the use of biological, neurological, or genetic tools and explanations in the behavioral and social sciences, while nurture advocates usually seek to preserve these sciences' explanatory and professional autonomy. The link is that reductionism versus emergence is also a debate about the relationship among sciences. The reductionist viewpoint tends to see the sciences as a nested hierarchy. The aspiration to reduce behavior to brains, neurons, and genes is also a bid to dissolve psychology and the social sciences into neuroscience, biology, and genetics. It is at least an effort to ensure the concepts and techniques of those more "fundamental" sciences play leading roles. In contrast, when Émile Durkheim defined sociology as the explanation of social facts with other social facts, or when early twentieth-century psychologists Kurt Koffka, Wolfgang Kohler, and Max Wertheimer understood mind, perception, and behavior as a holistic "gestalt" distinct from elemental parts and processes, they were arguing against reductionism and the hierarchical view of science. Different disciplines each have a self-sufficient mode of explanation, and thus they must be conceptually and methodologically autonomous from each other.

Behavior genetics was, of course, founded on the conviction that genes affect behavior and that the behavioral and social sciences have been almost corrupt in their resistance to biological explanation and their insistence on the sufficiency of socialization to explain and change behavior. As I showed in chapter 2, before the IQ controversy behavior geneticists sought to act on this conviction by creating an integrative disciplinary space that tried to sidestep the debate about reductionism. They attempted to do this, first, by creating a diverse research agenda in which reductionistic approaches to behavior existed alongside others and, second, by promoting a separate discipline where scientists could pursue these aims without necessarily assaulting the social and behavioral sciences' epistemic autonomy. However, with the sloughing off and withering of much the research agenda and the undermining of the field's disciplinary possibilities, behavior geneticists' reductionist convictions steered them directly into conflict with these sciences.

This chapter considers the period from the early 1980s to the late 1990s, which largely overlaps the time frame of the previous chapter. At this time behavior geneticists faced two basic problems in enacting their critique and spreading their scientific alternative. First, their main techniques for demonstrating genetic influences on behavior—twin and adoption studies and the estimation of heritability—had been criticized and undermined from seemingly every possible angle. Further, the revelation that Cyril Burt's data on the IQ of twins reared apart was fraudulent deprived the field of one of its most important datasets and cast wide doubt on the rest of its science. Second, their weak and stigmatized field provided them little support. The field's fragmentation cut off the psychologists, psychiatrists, and social scientists most likely to call themselves behavioral geneticists from the support of high status genetics and biology. The community was small and research funds were scarce. To defend themselves, behavior geneticists had congregated into a small set of institutions and adopted a bunker mentality about criticism. Even to some of its members, this attitude was stretching the limits of legitimate science. It may have sustained the community but failed to prevent their claims from being closely scrutinized and attacked, or shield their field from being denounced as eugenic, racist, and conservative science.

Facing these conditions, how could behavior geneticists promote genetics in psychology—the field from which most hailed—and other behavioral sciences? Sociological intuitions suggest that they would have to overcome their problems: strengthen and stabilize the field, recruit members and build institutions, draw boundaries and protect themselves from scrutiny, build consensus to quiet the controversy.[2] But instead behavior geneticists turned their weaknesses into strengths. Capitalizing on the field's flexibility and relative normlessness, they directed their attentions outward, seeking attention and recognition by provoking researchers in other disciplines. They extended their tools and ideas into a network of researchers but did not incorporate them into the field. Behavior geneticists made it possible to use their tools without having to join their field or assume their identity. They accumulated findings that gave behavior genetics an imprimatur of scientific success without addressing the fundamental critiques. Indeed, to this day the basic premises of behavior genetics' claims and techniques are still controversial across the sciences.[3]

This persistent dissensus about behavior genetics is one of the great puzzles about the field. Most sociological and historical accounts portray scientific controversy as a temporary, unstable condition. Controversy will erode a field's credibility if allowed to persist.[4] This chapter shows how behavior geneticists found ways to make dissensus, controversy, and provocation scientifically "profitable." Genetic reductionism, I show, has symbolic and practical value apart from "convincing people" and defeating the alternatives. The conflicts and provocations behavior geneticists engaged spurred practical scientific activity. Thus whether or not they were "right" or what they claimed was "true," the debates about their work kept the engine of scientific productivity turning. Misbehaving science can be made professionally beneficial.

This chapter describes different ways this provocative yet productive activity unfolded: through "hitting opponents over the head," accumulating a mountain of very similar results, and, more subtly, by "giving the field away" and recruiting others to engage such practices. The argument focuses on psychology, where behavior geneticists labored hardest to valorize their work. Finally, I consider how their aggressive stance drove behavior geneticists toward genetic determinist interpretations of their findings.

"Hitting Them over the Head"

The public side of behavior geneticists' bunker mentality was a pugnacious swagger. They saw themselves as targets of unfair, politicized attack. But they refused to be victims. BGA president Sandra Scarr compared behavior geneticists to freedom fighters and the students and radicals who protested them to Nazis.[5] She described a harrowing protest at an Arthur Jensen lecture where she threw a man who stormed the stage back into the audience. The next day they "formed a wedge of bodies" to protect Jensen from another onslaught. These were "mild stories compared to those others of you could tell," she said to the other BGA members. The clear message: behavior geneticists don't back down from fights. Many came to embrace the "bad boy" image they picked up during the years of combat. A reporter covering the famous Minnesota Study of Twins Reared Apart in 1992 quoted its director Thomas Bouchard, "'There are colleagues who think I'm a racist, sexist, fascist pig' . . . [he said] with his usual affable smile."[6] Several behavior geneticists echoed a view expressed by one: "There were some personalities in behavior genetics who deliberately enjoyed being iconoclasts and provoking others."[7]

This swaggering, aggressive disposition was more than a way to weather protests. It animated an approach to building the symbolic and material resources for securing scientific credibility and recognition, or scientific capital. For these behavior geneticists, the task was not to seek synthesis, integration, or sober rational persuasion, but to engage in polemical scientific attack, declaring themselves as crusaders who would rout the antigenetics heresy gripping behavioral science. One behavior geneticist described the milieu of his graduate program in psychology at the University of Texas at Austin, a center of behavior genetics, during the early 1980s: "It was the kind of world where you got tested for how intellectually tough and sort of fun you were. If you passed that test you were one of the boys. . . . Scientifically there was this kind of macho reductionistic attitude about everything, that old-fashioned soft psychology and the new hard sciences of brains and genes were just taking over. And that in twenty years psychology wouldn't exist anymore [it would just be] neuroscience."[8] Behavior geneticists saw their best defense as a strong offense, and so they were eager to challenge their scientific opponents. "They're not going to listen unless you hit them with a two-by-four. You just got to hit them over head with this stuff, be in their face," was many behavior geneticists' strategy at the time, according to one current field leader.[9]

The "hitting them over the head" style was common in publications and debates at this time. A particularly vivid example is the controversy that surrounded Sandra Scarr's 1991 presidential address to the Society for Research in Child Development.[10] Scarr argued that differences in parenting have very little impact on differences among children in their intelligence or personality, as long as that parenting was "in the normal range," not an extreme situation of abuse or neglect. She wrote, "environments most parents provide for their children have few *differential effects* on the offspring. Most families provide sufficiently supportive environments that children's individual genetic differences develop."[11] The evidence she cited was studies of twins and adoptees that argued monozygotic twins closely resemble each other and only slightly less so when they are raised in different families, and that adopted children raised in the same family don't resemble each other at all, nor do they resemble very much the parents who adopt them.[12] She claimed that across personality and intellectual traits, half or more of the differences among children are due to genetics. Of the remaining variance most of it is within families, not between families, which indicated that families are not making kids more alike. According to Scarr, the common idea that good families make their kids similarly smart while bad families make them similarly dull is not true. Scarr's account was that children's development is largely self-directed: Children gravitate toward environments and evoke responses from others that allow them to actualize whatever genetic capacities they're born with. For example, brainy kids respond well when parents read to them, or seek out intellectual stimulation if they don't. Shy kids avoid new people, and people encountering them usually respect their inhibitions. Scarr believed such self-directed processes happen except when a child is truly neglected, deprived, or abused. Thus "average expectable environments" and "good-enough parenting" will give children enough opportunities to find the environments that optimize their potential.

This was an extremely confrontational argument to developmental psychologists. It said, basically, parenting and, by implication, schooling and social interventions—that is, the things they spend all their time studying and trying to improve—don't really matter. Scarr charged that psychologists never figured this out because they spend all their time examining the detailed dynamics of socialization and never take a synthetic view using tools of genetic control. These claims drew heated responses.

Developmental psychologists Diana Baumrind and Jacquelyne Faye Jackson counterattacked in separate articles.[13] Both criticized behavior genetics methodologically, suggesting that twins and adoptees were not part of "normal" families and couldn't be generalized, that the methods were insensitive to gene/environment interactions and interactions are the essence of socialization, and that heritability estimates cannot ground claims that behavior is difficult to change. Both charged Scarr with failing to articulate what a "normal," "average expectable," "good enough" environment is, especially in light of the heterogeneity of parenting styles. Jackson discussed cultural differences between African Americans and whites, and Baumrind discussed international and other cultural variations within the United States, both of which strongly affect children's development. Further, Baumrind cited evidence about kids facing declining social conditions. Worsening education, slipping health, and rising violence and crime suggested that the "average" environment is not "good enough," she said. Both argued that research on the details and dynamics of socialization is crucial because it had been shown that different parenting styles and targeted interventions do change children's personalities and intellects. The essence of their arguments was that Scarr had run roughshod over the literature and exaggerated the salience of behavior genetics in the service of a pessimistic argument about parenting and change.

Rather than seeking rapprochement, Scarr's reply hit back even harder.[14] She targeted not only Baumrind and Jackson but also five other responses to her piece and broader objections to her perspective. She began, with some sarcasm, by noting that there was a lot of confusion about what she intended and said despite the brevity of the original text. She claimed her views follow logically from accepting an evolutionary view of humanity—a sly hint that her opponents reject a basic scientific truth. To charges that she is a "genetic determinist," Scarr said that "all scientists are determinists" making probabilistic statements.[15] Scarr's critics often invoked "gene/environment interactions" as an alternative to genetic determinism. Scarr dismissed this as "a descent from science into politically correct obfuscation. . . . Vitalists and Lamarckians are still with us."[16] She didn't address the critiques that she ignored normal cultural variations or used heritability inappropriately to discuss prospects for change; instead, she repeated several times that her claims were about the causes of individual differences (variance), not what cultures teach kids or how population averages come about. When on the attack, Scarr chopped down all of developmental psychology; on defense, she acted like all she had ever been talking about was sources of population variance—the distribution of differences around an average. This dodge slipped past the objections as if they were simply irrelevant. To the charge that her perspective ignores the detailed interactions of socialization, she dismissed the whole field out of hand: "Socialization studies . . . [are] hopelessly confounded," she said, because when parents raise their own children, it is not clear whether their parenting is effective or they have passed on genes that make their kids respond well to a particular pattern of parenting.[17] Further, she charged socialization researchers' "explanations of why and how siblings are similar and different in personality are ad hoc."[18]

"Hitting them over the head" was a strategy for building scientific capital that involved constructing one's intellectual interlocutors as mortal enemies and attacking them in spectacular, polemical fashion. Scarr's child development controversy exemplifies the strategy in vivid fashion. Her attacks were unrelenting; she admitted no failures and took no prisoners. She charged those who would criticize her as being not only unscientific but also adherents to laughably discredited ideas—vitalism and Lamarckianism—while turning scientific vices like "genetic determinism" into virtues.[19] Hitting them over the head thus meant raising behavior genetics' profile, not by synthesizing perspectives, resolving anomalies, or quietly settling disputes, but by fighting in the most garish, loud, unbending way possible. Behavior geneticists raised scientific capital by igniting and fueling controversy and conflict, not by ending controversies and convincing their detractors.

Accumulative Science

The confrontational style that behavior geneticists used didn't endear them to the psychologists and others they targeted. However, the conflicts had the effect of valorizing behavior geneticists' scientific tools. What makes this surprising is it occurred without behavior geneticists convincing their detractors—their research remained as controversial as ever. Doubts and conflicts did not undermine behavior genetics' credibility or scientific legitimacy. Instead, they spurred behavior geneticists to furious scientific activity and the repetition and accumulation of findings. Behavior geneticists did not convince their opponents, settle controversies, and resolve the critiques of their paradigm; instead, they buried their opponents under a pile of repetitive results.

Recall that in 1957, before the label "behavior genetics" had even been invented, the eminent psychologist Anne Anastasi called on her colleagues to go beyond "nature versus nurture" to ask *how* the two factors interact. Key to her argument was the idea that the heritability of numerous intellectual and personality traits had been demonstrated, and it was now time to move to the next phase. The early behavior geneticists saw heritability, if not exactly dead, as just a part of a broader agenda. The IQ controversy set in motion the disputes that moved heritability back to the center of the debate. Was the heritability of IQ about 0.8 as Arthur Jensen had argued, or, as Leon Kamin argued, did data problems mean that scientists shouldn't reject the null hypothesis, which he defined as meaning the heritability of IQ is 0? This dispute produced in the 1970s the flurry of estimates of IQ's heritability. At the field's founding heritability estimation had been declared dead; twenty-five years of controversy made it the central research task of behavior genetics.

Beyond the IQ debate, there were many technical challenges to behavior geneticists' methods that generated debate. For example, one way to estimate heritability is to compare how similar monozygotic (MZ) twins are to dizygotic (DZ) twins. For this to work one must assume that the environments they experience in a household are equal and those experienced when they're separated are truly different. Lewontin, Kamin, and Rose challenged both assumptions, arguing first that parents treat MZ, or "identical," twins more similarly—by giving them similar names, dressing them the same, and so forth—than do parents of fraternal, DZ twins.[20] They also charged that "separated" twins often grew up in similar households and sometimes were exposed to each other (living with relatives, for example). These arguments claimed an unknown portion of twins' similarity would be an environmental effect. Behavior geneticists developed a set of responses. One involved looking for cases where perceived zygosity was incorrect; Scarr and Louise Carter-Saltzman showed that twins incorrectly believed to be identical did not turn out to be more similar than other fraternal twins and argued that different treatment didn't violate crucial assumptions.[21] Bouchard and colleagues tested these assumptions by measuring aspects of twins' environments.[22] Behavior geneticists also argued that the greater similarity of MZ twins' environments did not undermine genetic findings; rather, they said that greater similarities were an effect of the twins' genes inducing them to seek out closer environments![23] Such clever intellectual jujitsu was a common gambit, but critics have charged behavior genetics with ad hoc reasoning and circular argumentation.[24]

These difficulties led some to favor studies of adoptees that work by comparing the correlations between adopted children and their adoptive parents, stepsiblings, and birth parents. But critics charged that adoption studies had their own problems, including grouping together children adopted at different ages with different degrees of exposure to natural family members, the selective placement of adoptees into environmentally similar families to their birth mothers, and that adoptees aren't representative of the population.[25] Behavior geneticists tried to show that inconsistencies and biases in samples of adoptees could be controlled and weren't too big of a problem.[26] Sandra Scarr turned the selective placement objection to her advantage: she estimated the effect of selective placement through the correlation of the adoptive child's birth mother with the household children she wasn't related to, then subtracted from this number the correlation between adoptive mother and adopted child to show that the environment has a vanishing influence on these children.[27]

Ostensibly, the stakes of these debates—which raged between the 1970s and 1990s and have not abated today—were the validity of behavior geneticists' techniques for estimating heritability. Behavior geneticists claimed that the existence of substantial heritability was a challenge to the idea of socialization. And if intelligence and behavior aren't much affected by social processes, then psychology might as well be reduced to biology. Politics entered as well: if behavior geneticists were right it seemed to confirm conservatives' pessimism about conditioning behavior for the better, if they were wrong, perhaps there was hope for liberal projects of social betterment.[28] The debates about behavior genetics were animated by this set of concerns.

The criticisms and behavior geneticists' responses—and there were dozens beyond those considered here—were arguable. But that was the point. As much as the substantive issues or the politics behind them, the debate was driven by the pragmatics of scientific work. The debates spurred the identification of new twin and adoptee populations, the administration of psychological evaluations, the development of statistical techniques, and the publication of new claims. The fact and the activity of the debate enhanced the legitimacy of behavior genetics in psychology. Much of the critique of behavior genetics was that it was racist, classist, or, at the very least, fatalist science that gave us false stories about human limits and the fixity of differences. By focusing on more specific empirical challenges, behavior geneticists pushed those political and moral critiques to the margins and focused attention on matters they found tractable. Much of the critique had been that heritability had no consequences for anything behavioral scientists really care about (that is, changing behaviors or understanding their causes). The well-known Thomas theorem, "If men define situations as real, they are real in their consequences," was in operation.[29] Whether or not heritability "objectively" mattered for psychological or social analysis, the fact of the debate confirmed that heritability was consequential.

An important branch of criticism had targeted the datasets behavior geneticists used. Leon Kamin attacked the major twins and adoptee datasets that had been the foundation of behavior geneticists' claims through the 1970s.[30] He argued that inconsistencies in the measurements of IQ, in the composition of the population samples, and in the evaluation of their environments should lead scientists to disregard any claims citing them as evidence. This, of course, spawned several debates, but such doubts also led behavior geneticists to generate newer, more carefully controlled datasets. Among the most prominent were the Texas Adoption Project (begun 1973), the Colorado Adoption Project (1976), the Minnesota Study of Twins Reared Apart (1979), and the Swedish Adoption/Twin Study on Aging or SATSA (1984).[31]

These projects asked twins and adoptees to fill out extensive questionnaires to gauge their cognitive abilities, personality traits, daily habits, likes and dislikes, social and political attitudes, and so forth. The Minnesota study, for example, invited pairs of twins who had grown up separately for fifty hours worth of testing. Scientists gave twins three different IQ tests, four personality assessments, three occupational interest tests, multiple physiological exams, psychiatric exams, assessments of religious and social views, and took oral histories of their life events, family background, education, and sexual activity.[32] Few studies were this extensive in their data collection. Most relied instead on extensive mailed questionnaires, but even these could run to dozens of pages. A relatively early and uncomplicated example was a 1963 survey of 850 twin pairs by the National Merit Scholarship Corporation. The survey ran to twenty-four pages, asked about respondents' activities, self-perception and personality, relationships, vocational and life aspirations, attitudes, health, even items in their rooms—it generated 1,610 variables in addition to, presumably, their college board scores.[33]

These new datasets opened up tremendous scientific opportunities for behavior geneticists. One part of this was the huge range of behaviors on which they could publish—from cognitive abilities, personality traits, and psychopathologies all the way to curiosities like television watching habits, political viewpoints, and beverages consumed. Table 5.1 gives just a taste of this range. As it turned out from their studies, identical twins almost always resemble each other a bit more than fraternal twins do, which means heritability can be calculated. This fact made twin surveys a protean resource that could be spun out in countless variations.

Why study all these different behaviors? Part of the reason was "hitting them over the head": continuing to provoke psychologists by showing that anything they might care about (and also anything they didn't) had a genetic component. If there was a behavior that some psychologist

TABLE 5.1. **A selection of significantly heritable traits**

Trait	Source
Intelligence	Bouchard and McGue (1981)
Personality	Loehlin and Nichols (1976); Bouchard (1994)
Schizophrenia	Gottesman and Shields (1976)
Aggressive behavior	Rushton et al. (1986)
Criminal behavior	Rowe (2002)
Educational attainment	Behrman and Taubman (1989)
Occupational status	Lichtenstein et al. (1992)
Job satisfaction	Arvey et al. (1994)
Homosexuality	Eckert et al. (1986)
Volunteering for combat	Lyons et al. (1993)
Divorce	McGue and Lykken (1992)
Church attendance	Truett et al. (1994)
Television watching	Plomin et al. (1990)
Conservatism and radicalism	Martin et al. (1986)
Attitudes (e.g., to jazz, nudist camps, and teen drivers, but not coeducation or straitjackets)	Tesser (1993)
Consuming alcohol, soda, milk, or coffee (but not diet soda or juice)	DeCastro (1993)

had developed an instrument to measure, then it was almost certain that some behavior geneticist would measure it in a set of twins or adoptees and offer a heritability estimate. Beyond the major personality and cognitive characteristics that psychologists really care about, the heritability of seemingly bizarre, mundane, or utterly culturally contingent behavioral traits seemed to proclaim with a flourish, "this too is genetic!" A figurative blow to the other side of the head. Beyond the effort to valorize genetic claims broadly, the crowding of the field encouraged the proliferation of claims. Some of these claims might be little more than curiosities scientifically, but they are also good candidates for attracting media attention (to which we turn in chapter 7).[34]

The number of traits behavior geneticists studied grew steadily, but the way they conceptualized behavior was very narrow. Behaviors were usually seen as static, relatively context-independent dispositions that could be measured with paper-and-pencil tests.[35] Sandra Scarr described this preferred view:

> The characteristics of real people in the real world *are* messy and confounded. . . . I think we are stuck with a synthetic view of such variables. . . . For studying the phenomena of people's lives, my preference is for global index variables over analytic, proximal ones. I recognize that this is a personal preference not shared by most psychologists. . . . I have been criticized by developmental psychologists . . . for not measuring the *processes* by which children come to be different from one another.[36]

The "global index variables" Scarr is referring to are things like scores on IQ, personality, psychopathology tests, and perhaps contextual variables like age, years of education, family economic status, and so forth.[37] Yet another front in the conflict: psychologists often criticized behavior geneticists as crude in these measures of behavior and inattentiveness to contexts in which behaviors happen.[38] There was no reason in principle why behavior geneticists couldn't have conceptualized behavior in more dynamic terms, studied it in experimental contexts, or chosen different versions of the same behaviors (for example, theories of moral reasoning or multiple intelligence instead of focusing on IQ). Indeed, some field founders and animal researchers favored more interactive and contextual accounts of behavior.[39] Behavior geneticists' narrow preferences seem to be due to the fact that in this period those advocating other conceptions of behavior were their enemies. These other views of behavior were pitched, in part, as critiques of behavior genetics and the biological reductionism it represented.

New datasets also enabled behavior geneticists to deepen the sophistication of their statistical tools. Behavior geneticists moved beyond the estimation of heritability in single traits and began to build and test models about the different ways genes and environment affect relationships among traits. Using their extensive new datasets, which often captured information about many family members and their environments, they could use statistical techniques like structural equation modeling to hypothesize different relationships and see which fit the data better.[40] For example, a system of family correlations might help a behavior geneticist test whether a parent's personality directly affects children's personality or whether the personality-to-personality effect is really due to the genes the parent has and has passed on to children.[41]

The most consequential of these developments were models that sought to slice up the environmental effects on behaviors. By comparing twins raised together and twins raised separately, behavior geneticists partitioned the environment into "shared" and "nonshared" components. This means the parts of the environment that siblings have in common, which is usually interpreted as parents' influence in the home, and parts that siblings don't, usually the idiosyncratic experiences they have in the world. A common finding was that genes and environment are each responsible for about half the behavioral variance. However, of the environmental variance most of it was from unshared sources and very little was from shared sources. As the leading behavior genetics textbook puts it, "children growing up in the same family [are] no more similar than children growing up in different families."[42] Many behavior geneticists have interpreted this brashly as: parents don't matter.[43] And they published—and conflicted with psychologists and others—many variations on this theme with different behaviors, family configurations, and models of the environment.

The proliferation of models and datasets, and the competition to publish, also led behavior geneticists to apply and interpret heritability in more flexible ways. With longitudinal datasets they looked at changes in the heritability of traits over the life course. (It usually goes up with age, which behavior geneticists took as another piece of evidence that people select into environments that fit their genetic makeup.[44]) Behavior geneticists also began to estimate the heritability of different subsamples in their datasets.[45] For example, Douglas Detterman, Lee Anne Thompson, and Robert Plomin argued in "Differences in Heritability across Groups Differing in Ability" that the heritability of very low IQ scores was higher than that of higher IQ; their possible explanations were either selection (that low IQ people participate less in research) or that low IQ people are different (have more highly penetrant genes, worse responses to environments, or less capacity to find improving conditions).[46]

Not everyone was happy with such efforts. One animal behavior geneticist I interviewed remarked on that article:

> And I gagged when I saw [the title of that article]. This is not possible, a title like that. Just the title is a conceptual error. Heritability is something that is specific to a certain population and certain characteristic. . . . There are two possibilities. Either high IQ and low IQ are two different characteristics that you both can measure in the same person like systolic and diastolic blood pressure, which I don't think they were intending to do. Or you're assuming that you're dealing with two different randomly mating populations that don't intermingle.[47]

His charge was that behavior geneticists were computing heritability scores willy-nilly without attending to the relevant conceptual limitations. "The computer program doesn't think. You can throw in anything, and it will render data."[48] As human behavior geneticists deployed their new datasets, the pressure to make novel claims encouraged them to relax or neglect some of the methodological rules that others believed to be absolutely essential. Thus the conflict between animal researchers (who tended to be methodological sticklers) and human researchers (who were more open in their interpretation of their tools) continued to simmer. However, the weakness of the field, particularly the norm against vigorous policing, meant that the conflict remained latent and the looser interpretations could flourish.

The philosopher of science Imre Lakatos described "progressive" research programs as those that grow, produce novel facts, refine their techniques, and generate increasingly precise predictions.[49] Thomas Kuhn defined cumulation as the progressive puzzle solving within a paradigm.[50] Behavior geneticists saw themselves in these terms. They cited the production of their findings as evidence—in study after study, behaviors turn out to be heritable and, almost always, nonshared environmental influences are greater than shared environment.[51] They saw their opponents in much of psychology and related sciences as faddish and cyclic in their ideas: in Lakatos's terms, "degenerating" programs that produced a "protective belt" of defensive ad hoc arguments that didn't address behavior genetic criticisms.[52] But this view was sustained by behavior geneticists' focus on the "tractable" empirical objections and ignoring the deeper theoretical objections.

Cumulation in behavior genetics was about "racking up" heritability estimates. This is a common critique of behavior genetics, and even some members of the field have commented upon it. One I interviewed said, "It's gotten a little harder, in fact it's gotten a lot harder . . . to publish. Fifteen years ago you could just routinely, you could get the you-name-it questionnaire, give it to a couple of hundred twins, compute the heritability, and publish it."[53] Meant to note the field's progress, this statement highlights accumulating heritability estimates as a long-standing self-justifying practice. In 1992 leaders of the SATSA project wrote, "Indeed it is a legitimate argument that the ubiquitous evidence for genetic influence on personality questionnaires makes it no longer interesting to document heritability for yet another personality trait."[54] Yet as critics Leon Kamin and Arthur Goldberger pointed out, SATSA leaders weren't persuaded by their own argument. In the next five years they used the data to publish a half-dozen other articles on the heritability of various personality traits.

To the extent that behavior genetics was progressive or cumulative in Lakatos's and Kuhn's terms, this was largely within their own research program. Behavior geneticists did not put the critiques of their paradigm to rest. They never "won" the argument by quelling the controversies about their science. These doubts did not fundamentally undermine their efforts because their research program was incredibly fruitful at producing variations on a theme. While behavior geneticists believed themselves to be engaged in progressive, cumulative science, for this to make sense we have to redefine these terms as the *accumulation* of similar findings. In this way permanent controversy was good to behavior geneticists because it created the conditions for them to build scientific capital with a mountain of findings.

"Giving the Field Away"

The pugnacious approach was not the only one behavior geneticists used to valorize their science in psychology and elsewhere. Some of them developed a more conciliatory style to attract research partners and extend the reach of their ideas. In the previous chapter, I quoted human researchers explaining their vision of behavior genetics as a set of techniques rather than a disciplinary specialty. These views were not only about a weakly institutionalized field but also their desire to share behavior genetics with other scholars. To quote a leading behavior geneticist:

> So, I think that's an important issue—[seeing behavior genetics as a] *tool not a school*—and giving the field away because it *makes the biggest impact*. And besides, there is so much to be done; the small group of people isn't going to do it anyway. . . .
>
> And, so by *getting people involved* in the research and *giving it away*, you get better work done because you get the *real experts* and people asking very driven questions that interest people in the field, unlike behavioral geneticists where they're usually viewed as carpetbaggers.[55]

The idea here is that behavior geneticists can do better by distributing rather than hoarding and defending their tools and expertise. This practice would help them overcome their limited numbers. More importantly it would help overcome their credibility problems: a perception that they were opportunistic "carpetbaggers" lacking the expertise and legitimacy to tackle certain problems.[56] Whereas Scarr's two-by-four approach seemed to elevate behavior genetics by demeaning the expertise of her targets, here "making the biggest impact" happened by cultivating allies, piggybacking on the authority of other experts, and creating new opportunities for accumulation. All of this increased behavior genetics' legitimacy, but it did not diminish its controversiality. Indeed, much of what made behavior genetics appealing was the capacity for provocation it gave its users.

The sociologist Nikolas Rose has used the term "generosity" to describe the field of psychology's distinctive path to professional power.[57] Most professions strive for power by exercising control: they limit access to their membership, expertise, and tools; closely regulate their members' conduct; and seek exclusive jurisdiction over domains of social life.[58] Canonical scientific fields, according to Bourdieu, are similar in that they tend to seek power by monopolizing and accumulating a specific kind of scientific capital.[59] Rose argues that psychology, in contrast, has gained professional power by "generously" dispersing its tools and ideas widely in society and being relatively unconcerned about defin-

ing legitimate and illegitimate ways of using them.[60] Generosity is an apt
characterization of the dominant knowledge production practices in behavior genetics. When behavior geneticists voice a preference for limited
association with a weak field—"I really try to grow with the problems"
rather than follow the "traditions or assumptions of any one club"[61] or
claiming "I don't want it to be a specialty . . . I want to give it away"[62]—
what they are arguing is that the power of behavior genetics lies in connecting its ideas and practices to those of other scientific fields, creating
new relations of exchange and intellectual hybrids that endow behavior
geneticists and scientists in other fields with new capacities. Generosity
in behavior genetics is at once more limited and more radical than in
psychology. On the one hand behavior genetics lacks applicability to social problems and the pervasive cultural authority of psychology (particularly its remarkable power to set the terms in which we conceptual-
ize ourselves). On the other hand, the generosity is more radical because
dominant behavior geneticists are not interested in building the field per
se, that is, strengthening the field's membership or institutions.

I have described elsewhere how behavior genetics has been received
differently in various fields with which its members have sought to establish exchanges.[63] Crucial to this reception are the historical associations
with the idea of "behavior genetics" and the kind of scientific distinction
it enables participants in different fields to claim. For human geneticists,
interest in analyzing behavior was outweighed by scientific doubts about
behavior genetics' methods and the race question, which made it relatively unappealing until the advent of molecular genetics (see next chapter). For psychiatrists, behavior genetics offered tools that aided the ascendance of the biological paradigm through the 1980s, but its distance
from clinical treatment and especially pharmacology limited its impact
in psychiatry. In genetics and psychiatry, the biological reduction of behavior is not controversial enough to be distinctive, and the controversies around epistemology and (racial) politics are not productive. In contrast, as I have shown, in psychology, behavior genetics has offered a
combination of ideas and tools that enable people to strike distinctive
positions while provoking fierce debate that stimulates practical scientific activity.[64]

What does the "gift" of behavior genetics offer its recipients? First,
it gives participants the opportunity to be productive in new terms; to
explore new data, questions, topics, and niches in their fields; to meet

new colleagues and seek their recognition; in short, it enables renewal
and movement, the pursuit of novelty and originality. This point is basic, but its importance cannot be overstated. In this vein behavior geneticists have occasionally won over those who had previously criticized
their work. "And I've found in the past, rather than arguing with people,
[I'll] say, 'Let's do a study,'" explained one prominent behavior geneticist.[65] In this way several highly regarded "environmentalist" psychologists—for example Jerome Kagan and Theodore Wachs—who had offered sharp critiques of behavior genetics work came to incorporate it
into their research, thus launching new career phases.[66] Certainly such
partnerships decreased disagreement among these scholars, but they did
not mean the resolution of the deeper divisions about behavior genetics
and, indeed, increased the potential for provocation in other ways.

Second, behavior genetics enables researchers to make clear and direct claims about human behavioral differences and to cut through much
of the complexity and qualification that usually accompanies expert discourse on this topic. A basic problem in social explanation is establishing
cause and effect. Say researchers observe a relationship between poverty
and low intelligence. Does the misfortune of poverty hinder people's intellectual development or does their low intelligence make them unable
to earn money? Behavior geneticists claim to have a solution based on
their faith that their research rests on causal bedrock. As the editors of
Behavioral Genetics in the Postgenomic Era write:

> DNA variation has a unique causal status in explaining behavior. When be-
> havior is correlated with anything else, the old adage applies that correla-
> tion does not imply causation. . . . Expression of genes can be altered, but
> the DNA sequence itself does not change. For this reason correlations be-
> tween DNA differences and behavioral differences can be interpreted caus-
> ally: DNA differences can cause the behavioral differences but not the other
> way around.[67]

This view also animates methods that don't directly measure DNA, such
as comparisons of twins and adoptees. Kinship relations set the degree
of genetic relatedness between individuals. Comparing IQ or personality
scores for differently related individuals who grew up with different degrees of contact with each other allows behavior geneticists to estimate
the degree of environmental and genetic influence on the trait's variabil-

ity. Because genes are the uncaused cause, the prime mover, behavior geneticists see their techniques as the way for the behavioral and social sciences to conquer fundamental epistemological problems.

Behavior genetics also offers researchers a powerful combination of "science" and "relevance." One behavior geneticist made a quip that could apply broadly in the human sciences, "nothing is more important to the field of psychology than to conceive of itself as a science."[68] What better way to be scientific than to study genetic influences (with their causal primacy)? Crucially, the "clinical orientation" of much behavior genetics, discussed last chapter, means that psychologists and social scientists can gain the association with genetics without having to retreat into the lab to stir test tubes or study fruit flies or mice. A big part of what makes behavior genetics distinctive and attractive to an array for researchers is the way it allows them to straddle symbolic divisions in the human sciences: "scientizing" the soft science side of psychology and "humanizing" the harder side.

Closely related to this prospect, behavior genetics offers researchers the opportunity to engage the pleasures of provocation. Many long-standing behavior geneticists obviously took pleasure in "hitting them over the head"; the new converts were often thrilled to pick up the old two-by-four. Several interviewees echoed this speaker's story:

> People who are zealous sort of on the environmental side—which was [prominent behavior geneticist] Tom Bouchard, he was an environmentalist. There's a bunch of psychiatrists who were Freudian that moved over to genetics, and they suddenly become rabid geneticists. You know everything is genetics . . . Well, [name of a prominent psychologist] almost became an embarrassment because he became so much into the genetics, but he was a psychoanalyst and a family systems person.[69]

In addition to being provocative among colleagues, through behavior genetics one can pursue public attention via the media. As scholars of science in the media have argued, genetic arguments about behavior have proven irresistible to journalists. Since the 1980s newspaper readers have been regularly greeted with reported discoveries of "genes for" diseases like heart disease and cancer but also behavioral traits like homosexuality, alcoholism, risk taking or thrill seeking, criminality and violence, schizophrenia, bipolar disorder, depression, and so on.[70] Linking one's research to behavior genetics is a chance to jump on this attention machine. University administrators and funders often reward this kind of publicity.[71] But this attention can be rewarding in itself; it provides social validation beyond the small circle of specialists.

Importantly, behavior genetics' generosity does not place overwhelming demands on its partners. Behavior geneticists kept their scientific tools portable and applicable to general problems in behavioral science. They "black boxed" many of their scientific tools and justifying tropes. Their partners did not need to retrain in biology and genetics; they did not need to enter the laboratory or understand how genes work; they did not need to become a member of the field by joining the BGA or adopting "behavior genetics" as a professional identity; indeed, they hardly needed to think about genetics at all. A field leader enthusiastic about these partnerships explained:

> You don't have to be a geneticist to do this stuff. These are tools just like a developmental psychologist uses. Longitudinal study methods, you can use genetic study methods as well. Do your same study, but do it in a genetically sensitive design, and look at how much more you'll get out of it. Just forget it was behavioral genetics. Just do your parent-offspring things and observing the home, but then after you're all done with what you were doing, I'll remind you that this was a [twin study].[72]

The behavior geneticist helps with the sample design—selecting and recruiting the genetically informative twins or adoptees—and statistical analysis. The partner supplies the substantive expertise in "asking very driven questions that interest people in the field." It is a strategy for exploiting comparative advantage that extends the reach of behavior genetics, and everything that comes with it.

"Giving the field away" was clearly more conciliatory and less aggressive than "hitting them over the head." But it wasn't about dissolving controversy or eliminating provocation. Rather than resolving criticisms and converting enemies, behavior geneticists targeted "people who were willing to be convinced."[73] Behavior geneticists gained allies and legitimacy. Their partners gained the means to enjoy the provocative potential of making genetics arguments. Everyone reaped the dividends of the practical activity of doing new scientific projects. This is not to say that these partnerships were made cynically but that the benefits they offered were distinct from the increments of increased knowledge they ostensibly delivered.

Provocation and Genetic Determinism

As we have seen, heritability estimates were central to human behavior geneticists' efforts to valorize their scientific capital. The different forms of provocation that behavior geneticists engaged helped commit them to their reductionist approach to psychology and related sciences and, in particular, to a genetic determinist interpretation of heritability estimates. The countless arguments critics made against heritability could be grouped into two basic sets. The first was that good estimates were impossible to get for human populations. The second was that even good estimates wouldn't yield useful, causal information about behavioral traits and prospects for changing them. This critique was akin to the old joke about the bad restaurant: the food is terrible and the portions are too small! The contradiction in the critique allowed the behavior geneticists to focus on the portion size—more and better estimates of heritability—while rarely addressing the question of whether it offered any intellectual nourishment.

Behavior geneticists couldn't leave the question of the value of heritability estimates completely unaddressed, but tackling it directly presented a basic dilemma. The interest and importance of heritability seemed to lie in its capacity to tell us about the cause of behavior and whether it can be changed. To make that argument would be clearly false in terms that were universally acknowledged. But to agree with critics that heritability gave no information about change would beg the question, why do this research at all? Behavior geneticists confronted this dilemma in a complicated way. On the one hand they aggressively defended a narrow definition of heritability to fend off their critics, but then they boldly extrapolated from it without acknowledging how this slipped beyond some core assumptions.

This can be seen clearly in the conclusion of David Rowe's *The Limits of Family Influence*, a behavior genetics classic that forcefully criticizes behavioral scientists' research on socialization by arguing that genes and nonshared environment have the greatest impact on differences in behavior.[74] In the conclusion, Rowe reviews the critics' charge that "the biological and the social are neither separable, nor antithetical, nor alternatives, but complementary." He responds that they are "confus[ing] individual differences [behavior geneticists' real target] with universal developmental processes."[75] Thus, as did Sandra Scarr, in the argument analyzed above, Rowe deflects the major criticisms by saying they're off point. Rowe concedes that the "malleability of heritable traits cannot be doubted," citing historical increases in height, a highly heritable trait, as evidence. However, he doesn't want to be seen as offering a "sugarcoating about how biological traits are not really determined."[76] He pointedly argues, "variation in shared rearing experiences is a weak source of trait variation. . . . Changes in parenting styles may make only a small dent in the sum total of our social problems. . . . If environmental interventions are to succeed, they must be truly novel ones, representing kinds of treatments that will be new to most populations."[77] It is worth noting, this is the same bold conclusion psychological behavior geneticists had been reaching since at least the Jensen controversy twenty years earlier.[78] The core of the claim is dependent on several elisions, however.

First, the idea that changing parenting could only make a "small dent in the sum total of our social problems" doesn't square with the claim that behavior genetics is only about the sources of individual differences. For Rowe's claim to be true, he implicitly has to be defining social problems as problems of variance: the problem with education is that we have a range of performance, the problem with crime is we have a range of propensity to bad behavior. The claim that family rearing is a "weak source of trait variation" and "changes . . . may make only a small dent" implies that the point of an intervention is to reduce variance. In some cases that might be true: we worry about inequality in school test performance, but rarely is the aim of education policy simply to reduce the spread of scores, getting more people away from extremes and close to the average. That would mean raising the bottom *and* lowering the top. Policy usually targets other possibilities, including raising the mean or moving the whole distribution; getting everyone above a threshold or minimizing the numbers at a lower extreme; or optimally matching individuals with services that will help them. While Rowe doesn't say anything false, the impact of the claim is based on implying that one very narrow definition of policy intervention stands in for all possibilities.

The idea that only a truly novel environmental intervention could matter is rooted in other elisions. The first part of this was that behavior geneticists believed their research covered the "normal range." They could acknowledge that families with twins or adoptees might not be "average," but they did come from a wide range of social environments and thus could well represent the "normal range."[79] This notion of representativity was reinforced by accumulation of very similar heritability estimates in many different samples of twins and adoptees—differences in background weren't changing the story about heritability and limited family influence. Other scholars would argue that behavior geneticists had slid over crucial details that would lead to much less deterministic view of the futility of intervention in the normal range. Psychologist Mike Stoolmiller showed that the fact that adoptions occur in families of higher socioeconomic status than average has biased the estimates of genetic and nonshared family influences dramatically upward. Correcting for the "range restriction"—the subset of the "normal range" that adoptive families represent—shows that "shared environment" or family influence, which behavior geneticists believed nonexistent, could account for up to 50 percent of variance in IQ.[80] These two versions of the representativity of behavior genetics research led to different conclusions about genetic determinism.

Another elision that led behavior geneticists to determinism was about the idea that only radical interventions could effect behavioral change. Behavior geneticists build an analogy to nutritional improvement and changes in highly heritable height. Once access to abundant calcium and protein is universal, only something like giving growth hormones will continue population improvements. Sandra Scarr's argument about "good enough" environments was something like this. Scarr had argued that the different "processes" of parenting were difficult to study on their own, their effects were confounded with genetics, and the full variety of them would be captured in behavior geneticists' samples anyway.[81] In other words, once parents supplied basic resources, nutrition, and security to their kids, differences in parenting wouldn't matter.

Historically, IQ scores have exhibited a secular increase not unlike that in height—the so-called Flynn Effect.[82] Economist William Dickens and psychologist James Flynn developed a model to address the paradox that IQ scores have risen despite high heritability. Their model builds on the idea of ubiquitous gene/environment correlations that Sandra Scarr described. Her interpretation was that genes determine the environments people choose, and therefore effects like parenting that we think of as environmental are actually genetic. Dickens and Flynn built a model based on similar assumptions about gene/environment correlations but came to opposite conclusions. First they showed several ways that the behavior genetic models statistically conceal environmental effects. Second, Scarr, Rowe, and other behavior geneticists had assumed that there were no environmental differences big enough to account for differences between people. Dickens and Flynn, however, showed that feedback loops could multiply and magnify small environmental differences into big performance differences—most kids have access to basketball courts, but slight differences in initial performance induce some to practice and others to give up, so the performance gap widens progressively. Interpreting the same behavior genetics data and concepts in different ways led Dickens and Flynn to a much less genetic determinist account of behavior.

In these two examples, non-behavior-genetics scholars used similar starting points and techniques available to behavior geneticists, yet they reached opposite conclusions about the relative power of genes and environment.[83] Behavior geneticists' polemical style of valorizing their research led them to plant deep stakes that were tightly clustered around a particular, basically genetic determinist, interpretation. Their consistent position, especially through the 1980s and mid-1990s, was that genes are the largest source of individual behavioral and intellectual differences, and that environmental sources are of the basically disorganized and unsystematic "unshared" variety. And, as in the Rowe example, the tendency was to interpret this as posing a tragic policy choice: fatalism about change or radical intervention, which would be costly and unpredictable. Non-behavior-geneticists, with the same information, emphasized larger and more varied roles for environmental influences and greater effectiveness for extant interventions. This is not to say that they were right, but that behavior geneticists' science did not determine their interpretations.

The suggestion that the same materials could be interpreted in less bold, reductionistic, or deterministic terms than did behavior geneticists does not mean they had some political ax to grind. It was due instead to what Bourdieu would call the space of intellectual possibility and the principles of vision and division through which behavior geneticists understood the meaning of their science. The structuring character of the space is well exemplified by the speaker quoted in the previous chapter whose lack of productivity in the same period almost cost him tenure because for a long time he couldn't envision "a path as someone who accepted some of the assumptions of the field but didn't accept a lot of the conclusions."[84] The practical logic of provocation as the means to produce scientific capital, not prior political commitments, made it nearly impossible to be a behavior geneticist who could mobilize the research in a nondeterministic, nonreductionistic way.

The broader point is this: "behavior genetics" was originally envisioned as the scientific field devoted to all aspects of the study of the genetic inheritance of behavior. But instead behavior geneticists represented a particular "progenes" position in different debates in the behavioral sciences, and many of their opponents were using similar substantive tools but not as "behavior geneticists." In an alternate universe, all these different viewpoints (and others) might have been debated within a much more intellectually heterogeneous behavior genetics. In this world, in the wake of the IQ controversy and the archipelagic restructuring of the field, the confrontational and provocative way behavior geneticists valorized their scientific capital led to their occupying a narrow, genetic determinist, intellectual territory.

Conclusion

In his recent book, sociologist Ari Adut has produced a theory of scandals, describing them as the disruptive publicity of transgression.[85] Adut highlights the strategic dimension of scandals—they occur when a party publicizes transgressions against some norm as a part of his or her own status enhancement strategies. In the typical political or financial scandal the publicizer is a journalist or regulator pointing out another party's transgression. But in others, for example political demonstrations or especially art scandals, the transgressor and the publicizer are the same person. The avant-garde artist gains infamy and status by wielding what art critic Robert Hughes called the "shock of the new."[86]

In this period behavior geneticists were acting something like transgressive artists. In the context of psychological science, behavior geneticists used genetic reasoning and reductionistic rhetoric to provoke the developmental paradigm about the dynamics of socialization. Such provocation, I have shown, became a crucial means for valorizing behavior genetics' scientific capital. Comparing science, where value is ostensibly grounded in truth, to art, where it is culturally arbitrary, is itself provocative. But behavior geneticists did not succeed in establishing their work as consensual truth; they were never able to settle the controversies surrounding their claims. Behavior geneticists found, much like artists, that provocation has its own value. There was a degree of awe, fear, and respect expressed when psychologists asked, "who are these maniacs?"[87] This translated into scientific attention, resources, and allies who could themselves enjoy the excitement of provocation. Scientific provocation departs from art in that it also functions through the pragmatics of scientific production. Cycles of claims and rebuttals generate grants, projects, data collection, presentations, and publications—the practical activity of scientific careers.

One result of this period of conflict was the curious valorization of behavior genetics' main research tools, the twin and adoption methods used to estimate heritability. As we saw in chapter 2, in some ways the scientific value of heritability was believed spent by the mid-1950s. Further, these tools have never stopped being controversial. Behavior geneticists have never convinced critics that heritability can be validly estimated for humans or even that heritability is meaningful to estimate. The story in this chapter is how behavior geneticists revalorized these tools by continuing to use them to accumulate results, addressing the critiques they could and ignoring other, deeper critiques. As publics get desensitized to scandals, so the field of psychology seemed to become desensitized to the critiques of behavior genetics. The activity of controversy has had an alchemical effect: the simple fact that twins resemble each other was transformed again and again into scientific gold.

The events here also show how a radical intellectual program of reductionism can have value despite having no chance of success on its own terms. Even the behavior geneticists admitted that genes didn't fully determine behavior and that biology couldn't capture everything about psychology. But they could use these ideas as wedges in their campaigns of provocation and recruitment. Further, psychology is a large intellectual space; there is room for the coexistence of warring perspectives. And, it should be noted, the disputes that behavior geneticists used to elevate themselves were not necessarily destructive to those they were attacking. These attacks too spawned practical scientific activity. Controversy could be mutually beneficial.

Finally, the relational dynamics of the field were important to the way provocation and controversy became embedded in behavior geneticists' practices. Part of this depended on behavior geneticists being largely dissociated from the disciplines of genetics and biology at this time. These fields were more concerned with mechanisms of development and less sanguine about heritability estimation; closer ties could have led to different emphases. More importantly, it is linked to the dominance of human researchers and their preferred vision of behavior genetics as a weak field—a set of positions and tools to enable members to compete in other fields. Animal researchers wanted behavior genetics to develop as an autonomous quasi-discipline with high barriers to entry and a vigorous competition among members for mutual recognition of their scientific activity. Had behavior genetics developed with this organization, scientific practices would have been less outwardly oriented and thus less dependent on provocation and controversy for valorization. Behavior genetics might have been less controversial and focused on different knowledge.

Such relational dynamics will play a strong role in the next chapter, which shows how behavior geneticists adapted to the advent of new molecular genetic technologies in the mid-1990s. As these technologies brought them into greater contact with molecular and psychiatric geneticists, behavior geneticists began to reposition themselves intellectually—as *environmental* researchers with a holistic concern for the *irreducible* individual. This reversal of fortunes shows that as much as intellectual commitments, genetic determinism and reductionism were also flexible tools for engaging an ongoing scientific competition.

CHAPTER SIX

From Behavior Genetics to Genomics

S cience today is driven by technological development. Genetics in particular has changed radically with the technologies developed over the last two decades that have given scientists tremendous access to the genome. Technological development is, to an extent, self-motivating. But as sociologists of science have often pointed out, scientists constantly look for technological fixes to diffuse and excuse the controversies of the present.[1] What about the quintessentially controversial science of behavior genetics? Would the dramatic technological advances of the molecular genetic and genomic era be able to pull behavior genetics from the quagmire of controversy?

The increasing availability of molecular technologies to study behavioral traits has been heralded with both excitement and anxiety. One of the basic criticisms of behavior genetics is that methods based on the analysis of variance using twin and family studies lack access to DNA. Its claims about the biological underpinnings of behavior are therefore unmoored speculation.[2] Molecular genetic developments since the 1980s have increasingly promised to make it feasible to measure DNA and correlate it to behavioral traits in humans. Even staunch critics of behavior genetics, like geneticist Richard Lewontin, had identified this capacity as crucial to judging the field a legitimate science.[3] Further, molecularization would improve behavior genetics by bringing in researchers with rigorous biological expertise and enabling integration with biological disciplines that had never been achieved. Unlike heritability, the basic interpretation on which scientists couldn't agree, DNA, with its unambiguous material nature, would provide a common starting point for research. Behavior geneticists have thus eagerly heralded molecular research as a "new era" that will "revolutionize" the field.[4]

On the other hand, some have viewed the molecularization of behavior with trepidation.[5] They worry that molecular genetics will reignite claims of the genetic determinism of behavior. If things were bad when we were talking metaphorically about genes via twins, imagine the potential for mischief when actual segments of DNA are involved. DNA correlated with behavior is still tricky to interpret. The association could be spurious, an artifact of the ways subjects are identified or behaviors are measured; simply identifying a gene or a chromosomal region doesn't say what the gene does or how it might influence behavior; genes always act in concert with many factors and usually have only a probabilistic influence on an outcome; and, as the example of PKU, a genetic metabolic disease treated by a dietary regime, shows, even a highly genetic trait might be changed by an environmental intervention. Critics feared that this highly technical, yet glamorous, science would be easy for journalists and the public to misinterpret. And as molecular genetics got more attention, became more costly, and scientific competition increased, scientists would have more incentive to make bold, agitating announcements.

This chapter shows that neither the hopes nor the fears were realized. Molecularization failed to quell the field's epistemic problems or quiet its controversies. But neither did it harden behavior geneticists' tendencies toward genetic determinism. Quite the opposite occurred. The older quantitative genetic claims of behavior genetics involving twins and family studies became a relative oasis of certainty and consensus as molecular genetics produced ambiguous results. And behavior geneticists became reluctant to make genetic reductionist claims about behavior and became much more interested in looking at *environmental* causes of behavior. Why?

There are two intertwined reasons for this counterintuitive outcome. The first is the dramatic failure of molecular technologies to fulfill anything close to the expectations that had been put on them. Molecular methods have proven limited at finding genetic correlates of behavior and terrible at meeting standards of replication. There have been similar problems across genetics, but behavioral traits, which are more difficult to define than heart disease or diabetes, have been particularly vexing. One might think that the technological quandary is sufficient explanation; after all, the problematic methods cannot quell controversy nor are they a sound basis for making deterministic claims. But a crucial factor was the field's social structure and competitive dynamics.

When the geneticists and psychiatric researchers bearing molecular methods entered behavior genetics, they reproduced rather than trans-formed its archipelagic structure. This means they formed or bolstered separate disciplinary islands rather than joining the preexisting behav-ior geneticists or working to build serious social connections among is-lands. Indeed, across islands there was a great deal of consensus about the value of molecular methods and especially about how to cope with their disappointing results, but this did not overcome the competitive dynamics they spurred. The arrival of molecular genetics presented the veteran behavior geneticists with a dilemma. The technology presented an obvious opportunity to explore new research questions and collabo-rations. Yet competition with the new contingent of molecular genetic hotshots also threatened to devalue their expertise and scientific invest-ments. Thus in addition to taking on molecular genetics they also sought to distinguish themselves. They did this by reimagining themselves and their research in environmentalist terms while quietly disparaging mo-lecular geneticists as crude reductionists. Particularly noteworthy is that behavior geneticists repositioned themselves in almost precisely the same terms they had disparaged in their competition with psychologists and social scientists that we considered in the previous chapter.

This chapter considers the molecular era in behavior genetics—roughly the mid-1980s to the present. I write in the past tense, but prob-lems of molecular genetics are still unresolved and continue to be de-bated today. It is important to note that this chapter focuses on the collective problem of coping with (doubtful) new technology. I defer un-til the next chapter consideration of the many problems of public contro-versy that accompanied molecular genetics, especially the key issues of public communication, media hype, and the resurgence of race differ-ences claims. This chapter proceeds by considering first the outsized ex-pectations for molecular genetics of behavior and then the ongoing sci-entific failure to meet them. I then discuss how scientists have practically and ideologically coped with these disappointments, and finally how be-havior geneticists adapted to and were transformed by the competition with molecular geneticists.

High Hopes for Molecular Genetics

The late 1980s and early '90s were an exciting time in genetics. The in-vention of new techniques and technologies for manipulating DNA were opening up tremendous possibilities for research. With new tools the amplification and sequencing of DNA became quicker and cheaper, which made knowing the sequence of organisms' whole genomes possible. The identification of markers across the human genome made it possible to locate genes associated with different traits. And it became possible to directly manipulate animal genomes, inserting or "knocking out" particular genes to observe their effects on development and behavior.

Genetics was growing, attracting money and talent. But geneticists' ambitions were vast. They envisioned sequencing the entire human genome, the Human Genome Project (HGP); now technically possible but still tremendously costly in time, money, and labor. Genetics leaders needed to convince the public (and sometimes each other) that it was worthwhile to spend $3 billion over an estimated fifteen years.[6] So they hit the pavement with a public relations campaign to rally support.

That campaign had two basic parts: revolutionizing our understanding of human disease and unlocking the secrets of human nature, which is to say, behavior. In his humbly titled essay, "A Vision of the Grail," the eminent geneticist Walter Gilbert argued that the HGP will tell us, "What makes us human?"[7] James Watson, codiscoverer of DNA's structure and first director of the HGP, often proclaimed, "we used to think our fate was in the stars, now we know it to be in our genes."[8] The "horrors of the deranged mind," he also stated, "more than give us reason to find the genes we know are there."[9] Biochemist Daniel Koshland, editor of *Science*, frequently used his editorial page to champion the HGP and to argue that its impact would be tremendous on crime, homelessness, and other social problems related to mental illness. Social welfare and public safety programs, he claimed, are Band-Aid solutions that fail to address the real, genetic roots of social problems.[10]

For some geneticists the appeal of studying behavior was undeniable. Accustomed to working on publicly inscrutable molecular mechanisms, behavior genetics offered them the opportunity to attract public and media attention and present themselves as authorities on "what makes people tick," as Dean Hamer, father of the "gay gene," put it.[11] The words of Watson, Koshland, and others were more than a sales pitch; they were a road map for geneticists to insert themselves into the center of crucial public issues, with therapeutics, not eugenics, being the aim. President George H. W. Bush declared the 1990s the "Decade of the Brain" to raise the public profile of biobehavioral research. An estimated 26.2 percent of Americans suffer from a diagnosable mental illness each year.[12]

Thus behavior, and in particular mental illness behaviors, seemed an important way to demonstrate the dramatic potential of genetics to medicine and human welfare. The field was economically enticing, too; this was also the era of the boom in blockbuster pharmaceuticals for depression (Prozac entered the US market in 1987 and Zoloft in 1991), biotech startups, and the millionaire scientist.

These conditions also attracted more psychiatric researchers to genetics.[13] Since the 1970s, biology had become ascendant over psychodynamic and Freudian approaches in psychiatry.[14] Psychiatric and behavior genetics, using twins and adoptees to demonstrate the heritability of schizophrenia, had played a role in this transition, but psychiatric researchers often viewed it as less than cutting edge. One psychiatric geneticist I interviewed said that when he began his career he was told: "You've got no future in psychiatric genetics. You know, they've done twin studies; they've shown there is a genetic contribution to schizophrenia. That's it. You know, where else can they take it?"[15] Molecular genetics, with its promise of pointing to DNA and the biochemistry of mental illness, was another matter altogether. As the psychiatric geneticist Kenneth Kendler has reflected:

> [There was] rapid, and nearly relentless, success of human genetics in mapping the classic Mendelian human genetic disorders. Few in the psychiatric genetics community could avoid feeling envious as these major disorders were mapped one by one. There was a great desire to get into the line with the hope that schizophrenia, manic-depressive illness, panic disorder, or alcoholism would be next.[16]

Molecular genetics was an opportunity for psychiatrists, always low on the prestige hierarchy of medical science, to align themselves with the most important and dynamic scientific developments of the day.

This burgeoning of interest in behavior genetics brought a rush of resources and personnel to the field. Research funding poured into the field. Figure 6.1 shows the number of grants made by the National Institutes of Health listing "behavioral genetics" as a keyword. From the mid-1970s to the mid-1980s there was a modest, but declining, number of grants being funded. From 1985 to 1990, only a handful of grants were made, but, starting in 1991—the era of molecular genetics—NIH rapidly increased grant making.[17] These trends are corroborated in data on the amounts of money coming into the field, as figure 6.2 shows. Grants for

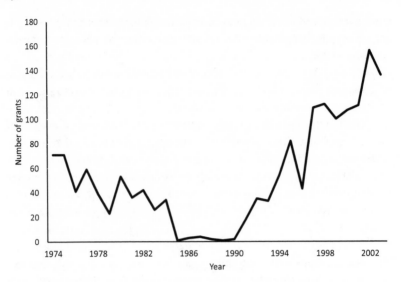

FIGURE 6.1. NIH research grants in "behavioral genetics" (1974–2003)

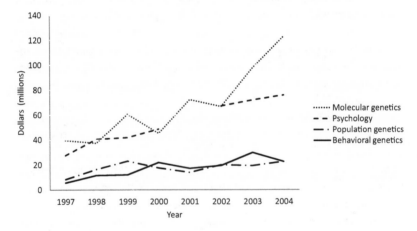

FIGURE 6.2. NIH funding for "behavioral genetics" and other fields (1997–2004)

"behavioral genetics" grew from 1997 to 2004—the average annual total was about $18 million during this period. By way of comparison, be-
havior genetics was funded much less than the more general disciplines
of molecular genetics and psychology but at almost the same level as the
much more established subfield of population genetics. The cumulative

amount of money spent on new grants for behavior genetics from 1997 to 2005 was over $600 million for 558 separate grants to 491 different principal investigators.[18]

This expansion in funding support was accompanied by an expansion of scientific institutions as well. In 1992 psychiatric geneticists founded the International Society for Psychiatric Genetics, and it sponsored the *American Journal of Medical Genetics Part B: Neuropsychiatric Genetics*. In 1996 a group of animal behavior geneticists and genetics-oriented neuroscientists formed the International Behavioural and Neural Genetics Society, and in 2001 it launched its own journal, *Genes, Brain, and Behavior*.[19] Several other journals were founded around this time, including *Psychiatric Genetics* (1990), *Molecular Psychiatry* (1996), and *Neurogenetics* (1997). And, of course, the Behavior Genetics Association continued on.

This growth brought many new scientists to the field, though precise numbers are unavailable. In 1985 the BGA had 388 members, and this swelled to 578 in 1996.[20] The combined membership of the three main societies in the late 1990s was in the neighborhood of 1,000.[21] The nearly 500 principal investigators of funded grants offer another estimate, since each of them is likely to lead a group of several other researchers (junior colleagues, technicians, postdocs, and graduate students); perhaps 1,500 to 2,000 researchers were working in behavior genetics from the late 1990s to early 2000s. Thus, though behavior genetics had grown rapidly during this period, it was still a small- to medium-size field.

The new entrants to the field had different disciplinary backgrounds and expertise—more medical and molecular—than the long-standing behavior geneticists. While this certainly created tensions, which I consider later, field veterans had access the new research tools. Some certainly retrained, but most were able to collaborate with molecular genetics experts. In many ways the "molecular revolution" in genetics was not conceptual but technological. Many experimental processes were black boxed and automated; that is, they could be deployed without engaging their technical complexity or ambiguities. Less and less was meaningful work necessarily tied to "wet lab" expertise, which could often be outsourced. The expertise of behavior genetics veterans was not fundamentally undermined, because these technologies focused much of the labor of research on study design, recruitment and collection, data coordination, and statistical analysis. These tasks were continuous with those behavior geneticists had always done, and, indeed, they could broker access to valuable data and populations.

Put differently, mutually low barriers to entry aided the influx and mixing of new researchers. Behavior geneticists had left the field relatively undisciplined with intentionally porous boundaries, requiring little of new scientists interested in participating. And this was matched on the other side by molecular geneticists' eagerness to work on questions of human nature and the black boxing and portability of their technologies that made them relatively easy to apply and widely available.

Failures of Molecular Genetics of Behavior

Molecular genetics has been a major disappointment, if not an outright failure, in behavior genetics. Scientists have made many bold claims about genes for behavioral traits or mental disorders only to later retract them or to have them not replicated by other scientists. Further, the findings that have been confirmed, or not yet falsified, have been few, far between, and small in magnitude. Molecular studies of behavior have often been inspired, on the one side, by universally high estimates of heritability for behaviors and, on the other, by the identification of genetic mutations for certain forms of mental retardation (like Fragile X and PKU). But there has been a tremendous mismatch between these successes and the limited achievements of molecular studies of behavior. Here are some prominent examples of failure.

An early finding heralding the promise of the molecular era was Janice Egeland's 1987 announcement of a genetic linkage to manic depression found by analyzing the pedigrees of Amish families.[22] But just two years later the finding was retracted when a member of the family from a different branch of the family tree was diagnosed, which rendered statistically insignificant the link between the putative chromosome location and disease in the family line.[23] When the finding was reported in a conference the eminent molecular geneticist David Baltimore remarked, "as an average reader of *Nature*, what am I to believe?"[24] It would get worse: reviewing nineteen other failed bipolar disorder linkage claims in 1996, geneticists Neil Risch and David Botstein wrote of the "euphoria of linkage findings [is] being replaced by the dysphoria of non-replication . . . creating a roller coaster–type existence" for field members and observers.[25] Molecular genetics research on manic depression remains unsettled today.[26]

In 1993 cancer geneticist Dean Hamer's group announced, to tremendous public interest (both fascinated and outraged), a link between a region on the X chromosome and male sexual orientation in the population he studied.[27] Academics and the lay public heatedly debated the scientific and social meaning of the claim.[28] In an unusual move, Hamer cowrote a popular book, *The Science of Desire*, detailing his process of discovery.[29] But another lab's attempt to replicate the finding on an independent sample failed.[30] Scientifically, this claim has largely faded away, though it lives on in the public imagination.

Another focal point for molecular genetic research has been alcoholism and alcohol use. Much attention was focused on research by Ernest Noble and Kenneth Blum, who claimed to find an association between alcoholism and variants of genes that produce a receptor for the neurotransmitter dopamine.[31] This suggested not only a genetic link but also a pathway for uncovering a molecular mechanism involving the brain's ability to produce "pleasure." Journalist Constance Holden reported the history of failed replications and even the growing annoyance of Noble and Blum's colleagues that they continued to forcefully press their case.[32] Sociologists Peter Conrad and Dana Weinberg put this research in a broader context, arguing that media coverage had given the impression that the "alcoholism gene" had been discovered three times by 1996.[33] Further, a small set of researchers had trumpeted their supposed success repeatedly with little reference to the unsettled state of knowledge.

One of the most celebrated recent behavior genetic studies was led by Avshalom Caspi and Terrie Moffitt about a gene/environment interaction in depression.[34] The researchers found that the effect of stressful life events on depression is mediated by differences in the promoter region of the serotonin transporter gene. People experiencing stressful life events are much more likely to become depressed if they have one or two copies of the "short" allele compared to those with two copies of the "long" allele. *Science* mentioned it among the most important scientific discoveries of 2003.[35] Attempts to replicate the study have been mixed. In 2009, however, Kathleen Merikangas and Neil Risch pooled the data from fourteen studies into a meta-analysis that found no effect of the gene regardless of life experience.[36] Epidemiologist Stanley Zammit and colleagues have looked at the recent trend of studying gene/environment interactions (as in the Caspi study) in behavioral traits. They argue that claims of replication haven't been warranted and, indeed, that the whole endeavor is of limited value to understanding the cause and treatment of mental disease.[37] This promising line of research, widely attractive for focusing on interaction rather than determination, has also become derailed.

Studies have also failed to identify robust links between DNA and cognition/intelligence despite it being one of the traits consistently showing the highest heritability estimates.[38] A recent study published in *Molecular Psychiatry* (involving thirty-two coauthors) used the latest genome-wide association technology to scan the genomes of 3,511 people to look for differences associated with cognitive traits, including IQ.[39] The study turned up no "IQ genes," and together the 550,000 "single nucleotide polymorphisms" (points where the genome is known to vary by one genetic "letter") explained only about 1 percent of the population variance in cognitive ability. Analyzing the data different ways confirmed the heritability estimates that low-tech twin studies provide. Although not exactly a "failure," this publication represents a tremendous amount of money and effort for an extremely limited success. But as anthropologist Jonathan Marks put it, "In what other field can you purport to explain 1% of a phenomenon (and even that 1% is correlational, not causal), and still get headlines?"[40]

These examples are part of a broader pattern of disappointing results. In 2002, geneticist Joel Hirschhorn and his colleagues performed an exhaustive review of genetic association studies.[41] Of the 600+ reported associations to any medical trait (twenty were psychiatric), 166 had been studied three or more times and only six had been consistently replicated. This was an early entry in what has become a parallel literature reporting molecular genetic disappointments and offering instructions for cautious interpretation and improvement of the corpus.[42] By this point a secondary backlash literature has emerged charging the field with being now too stringent and sowing doubt with "pseudo-replications."[43]

The growing disappointment of molecular genetics in behavioral studies has been an ongoing process. There have been many sharp pangs, when a particularly celebrated claim has been knocked down, but this has mostly been an unfolding drama without decisive moments. It is important to note that the disappointments or failures of the field are by the practitioners' own standards. There have been additional critiques, for example, of the decontextualized definition of behaviors that are historically and culturally specific. Critics have also disputed the usefulness of searching out genes without a model of the organism's development.[44] But the problems here have been in the molecular researchers' own terms: the replicability of claims and the statistical significance and magnitude of effects. This scientific failure has been a well-kept, open secret.

Coping with Failure

Leading psychiatric geneticist Elliot Gershon said in 1994 that the field "could not survive" a set of disconfirmations like those of the late 1980s.[45] He was wrong both in his optimism that the hard times were over and also in claiming they would be fatal. How did behavior geneticists manage the disappointments of molecular genetics? Some critics would call for the cessation of research.[46] During the earlier IQ controversy, crisis spurred many to flee the field. But in the molecular era, behavior geneticists produced a set of arguments and practices that pushed problems down the road while expanding their research opportunities.

One of the most basic strategies for dealing with the disappointments of molecular genetics has been to lower expectations. In the early twentieth-century prehistory of behavior genetics, it was common for scientists to assume that complex behavioral traits—feeblemindedness, pauperism, seafaring-ness—were inherited wholesale in basically Mendelian fashion.[47] Geneticists had long known this view to be false, and indeed, the application of molecular genetics was predicated on its falseness. If the patterns were Mendelian, it would have been possible to discern them through statistical analysis of family data (though not identify their location on DNA). Still, through much of the 1990s at least, there was great hope in behavioral and psychiatric genetics that genes of relatively large effect would be found for behaviors and mental disorders. "It is obvious that these are likely to be forthcoming very soon," wrote Michael Rutter and Robert Plomin in 1997.[48]

As time has gone on, behavior geneticists have come to expect smaller and smaller effects. As several field leaders wrote in their introduction to the volume *Behavioral Genetics in the Postgenomic Era*:

> Perhaps 1 gene will be found that accounts for 5% of the variance, 5 other genes might each account for 2% of the variance, and 10 other genes might each account for 1% of the variance. . . . Not long ago, a 10% effect size was thought to be small, at least from the single-gene perspective in which the ef-

fect size was essentially 100%. However, for behavioral disorders and dimensions, a 10% effect size may turn out to be a very large effect.[49]

In 2003, when this was written, it would have been seen as a bold, yet mature, admission. In many cases, it might take sixteen different genes to account for only a quarter of the population variance in a behavioral trait. The field is confident enough, they seem to be saying, that this isn't a problem. But by 2011, when 500,000 differences across the genome could account for only 1 percent of the variance in IQ, the humble picture from 2003 suddenly looked wildly optimistic, and behavior geneticists have had to revise expectations further downward.

Another strategy for coping with these failures has been to reconceptualize behaviors. One version of this has been to argue that many behavioral traits and mental disorders that look like a single thing at the population level are actually agglomerations of individuals expressing different causal pathways.[50] Robert Plomin, Michael Owen, and Peter McGuffin discuss this with respect to mental retardation.[51] Some individuals diagnosed with mental retardation have the single gene metabolic disorder PKU, others have various forms of Fragile X, and others have complex combinations of genes working in probabilistic fashion with the environment. The idea here is that "mental retardation" is actually comprised of different subtypes, some genetic, some environmental, some complex combinations of both. A similar logic might apply to traits like intelligence; from a genetic point of view there may be many different types of, say, high intelligence that produce similar scores on an IQ test. Thus genes still matter, but they are difficult to identify in scans of aggregated populations.

In a related move, traits have also been reconceptualized as collections of "endophenotypes."[52] Psychiatrists define mental disorders as aggregates of symptoms. As Irving Gottesman and Todd Gould describe, those diagnosed with schizophrenia might be exhibiting, among other things, poor working memory, eye-tracking dysfunction, difficulty filtering information, and exaggerated startle responses. Rather than investigating the genetics of "schizophrenia" writ large, the endophenotype strategy would investigate the genetic links to such intermediate phenotypes. The behavior reconceptualization strategy suggests that molecular genetics' problems are due to aggregating too many different kinds of people into one category. The endophenotype strategy is about aggregating too many separate traits under the same label. If these strategies for confronting the failure of molecular genetics work, then they will be accompanied by a dissolution or disaggregation of current behavior and disorder categories.

Technological optimism is another coping strategy. One skeptical behavior geneticist told me, "instead of ever facing up to what hasn't worked, the field always just moves on optimistically to the next level of technology which is always available."[53] Behavior geneticists have tried to escape failure by climbing a technological ladder. At the bottom, linkage analysis looks at family pedigrees to identify the genetic markers or chromosomal regions that affected family members hold in common. The technique requires few markers to scan the entire genome but has poor resolution for detecting small genetic effects. When linkage faltered, researchers climbed a rung and sought behavioral associations with candidate genes. This technique enables researchers to ignore family relationships and compare individuals who have a trait and controls who don't, but markers must be very close to the "candidate gene" of effect to be detected. Researchers also realized that the candidates could be wrong, and they needed to look for unknown associations. Thus, one rung higher, researchers have turned to genome-wide association studies (GWAS) that scan 500,000 to 1,000,000 differences in single nucleotides across the entire genome. As this tool has also been disappointing, geneticists have begun to look up the ladder at whole exome sequencing (which would yield the entire part of the genome that is transcribed and translated into proteins), whole genome sequencing (which would capture all types of genetic variation), or gene-expression or epigenetic arrays (which reveal how genes are actually expressed or the inheritable changes that can regulate expression).[54] This optimism that the next level of technology will overcome past disappointments is accompanied by doubling-down and scaling-up. That is, the basic diagnosis of failure has been that researchers have looked at too few points and types of genetic variation in too few people. The response is therefore that researchers need to invest more money in higher-resolution technologies and to collect larger samples of research subjects. Recently, ninety-six psychiatric geneticists signed a collective letter asking funders not to "give up on GWAS" despite the lack of promising results.[55] They claimed that if they had funding to quadruple sample sizes, psychiatry's genetic findings would match those now available for many non–mental illnesses and traits. As a result, behavior genetics, but particularly psychiatric genetics, is becoming more like "big science" where research groups from several institutions must pool their resources (especially their research subjects) to generate publishable results with the labor and the credit distributed among dozens of coauthors.

Sociologists Michael Arribas-Ayllon, Andrew Bartlett, and Katie Featherstone have identified a key rhetorical strategy that psychiatric geneticists use to cope with disappointing research.[56] They show that in review articles that appraise progress in psychiatric genetics, authors universally mobilize a discourse of "complexity" to account for the field's disappointments and failures. Complexity in these accounts refers to the "witches' brew" of multiple small genetic effects, gene interactions, gene/environment interactions, and nongenetic effects that are imagined to cause mental illnesses. They show that this discourse "incorporates criticisms, while at the same time deleting or minimizing the controversies from which they are derived."[57] Complexity talk allows the field to exonerate failure, justify intensified efforts, defend against charges of determinism, and project moderation and responsibility.[58] What these authors don't point out is that a skeptical reading might see complexity talk as little more than cant: perhaps it is less a positive description of how causal factors fit together than an admission of ignorance based on the failure of simple explanations. Even so, this language serves the important rhetorical function of justifying and integrating the field's practical adaptations to keep activity churning along.

An interesting feature of these failures and coping strategies is that the elements of disappointment are discussed constantly, but there has been surprisingly little rumination on the meaning of their accumulation or what would constitute general "failure."[59] One of the few behavior geneticists vocal in criticism said, "if there's any negative feeling about me in the organization it's that people feel like things that I've written make it harder for people to get grants. Because they don't want someone saying, 'Well the truth is we're not making any progress . . . there's no big answer out there.' . . . That is not what the funding agencies want to hear."[60] At least ninety-six researchers seem to be worried about exactly that. This is a different kind of bunker mentality, one based less on defense against outside attack and more on the collective censorship that Bourdieu argues is based on intense investment in common intellectual struggle.[61]

Beyond these arguments and practices for coping with molecular genetics' scientific problems, how was the field changed by the competition the technology introduced between veteran behavior geneticists and the newcomers?

Reproduction of the Archipelago

One of the original aspirations of advocates of molecular research in behavior genetics was that it would help integrate the field around a common paradigm.[62] On top of complexity talk, Arribas-Ayllon and colleagues also showed that a crucial response of researchers to accumulating disappointments has been to call for greater interdisciplinary exchange.[63] In health and sickness, then, the move to molecular genetics is motivated partly by a yearning for greater union.

Molecular genetics, as we saw, brought resources and growth to behavior genetics, but they were incorporated mostly by reinforcing the field's archipelagic structure. The psychiatrists, geneticists, and neuroscientists who took an interest in behavior genetics didn't typically join the BGA or publish in *Behavior Genetics*; they formed new societies and journals. They sorted into distinct, discipline-based islands. They were largely interested in engaging those on the same island and seeking the recognition of their home-discipline mainlands. Engaging those on other islands was a much lower priority. While interest was high in researching the genetics of behavior, this didn't mean the new entrants were interested in *joining* the field's prior inhabitants or taking on the behavior genetics *identity*.[64]

Indeed, substantial tensions accompanied the influx of new researchers to behavior genetics. The veteran behavior geneticists and the new entrants were often wary of each other and eager to distinguish themselves. Psychiatric, behavior, and molecular geneticists often disparaged each other with what Freud called the narcissism of small differences. Despite differences in emphasis, they all worked in the same intellectual paradigm, used largely the same suite of scientific tools, cited each other's work, and even collaborated, yet were loathe to be to closely identified. A molecular geneticist I interviewed put the point bluntly:

I mean a big problem here is trying to map out a separate domain of behavior genetics, because there is no such separate domain. I mean there are a few people who are geneticists and behaviorists and who don't do other things, but they don't really make breakthroughs, you know. I mean, if you think that it's a breakthrough to characterize the heritability of a behavior, then fine. But if the issue is, "what is the origin of that behavior?" then elucidating that is the achievement of different methodologies, including studies on functional gene variation and measurement of phenotypes that are in the pathway of that gene and its relationship to behavior.[65]

The speaker is utterly dismissive of the scientific contributions of those who call themselves "behavior geneticists" and rejects the suggestion that he be counted among their ranks. Insofar as "behavior genetics" is a bona fide scientific field, it is one that doesn't make "breakthroughs." Insofar as one aims to make breakthroughs about the relationship between genes and behavior, one must do it as a part-time interest of a deeper expertise in molecular genetics.

The veteran behavior geneticists were equally wary of these newcomers. The major concern was molecular geneticists' focus on the lowest levels of biochemistry. One behavior geneticist put his misgivings this way:

I just wish people could be more tolerant of other levels of analysis—different questions, you get different answers. But there still, especially in molecular biology, there is such a strong vibe towards the cellular, you know, reductionist approach. And the rest is not really viewed as science. . . . [Name of a famous molecular geneticist is] one who has that need to criticize me saying, "you know unless you get down there, you know, at a single gene and understand how it works at a cellular level, you're not doing science." . . . This is "social work" I think is what he calls the rest of it.[66]

Thus behavior geneticists were concerned not only with the reductionist views of molecular geneticists but also with the arrogance they perceived: molecular geneticists elbowing their way into the field and redefining science in a particular way. This is, of course, an ironic role reversal, since in the 1980s and 1990s psychologists and social scientists viewed behavior geneticists in almost the exact same way.

Animal behavior geneticists also had misgivings about molecular geneticists. One mouse researcher complained:

You see that there is a complete ignorance of anything that was before. Only a few years ago I read a comment, I think it was in *Science*, where someone

was writing apparently without blushing or even hesitating, that [it has recently become possible to] do behavior genetics with mammals because now
we can do knockouts and transgenics.[67] And I thought, wait a second. . . .
There's a lot of animal behavior genetics around and so you see that sometimes people reinvent the wheel and start doing the same things again.[68]

This speaker resented the way the new researchers were rushing into the
field, presuming it to be virgin territory, and blithely laying claim. They
saw an utter disregard for their decades of work and their specific expertise melding genetic methods to the vagaries of animal behavior. Another example: at a summer course on behavior genetics I attended, a
lecturer related a story about some molecular genetics newcomers to the
field. They claimed to have found a gene accounting for defective memory in mice. Behavior geneticists trying to replicate the work realized
that the poor memory performance was an artifact of the innate blindness of the strain of mice that had been used. The lesson of the lecture
was to look for confounding factors in behavioral measures. But the sub-
text was frustration that invading molecular geneticists have been heed-
less of the expertise of behavior geneticists.[69]

Molecular researchers' move into the realm of behavior was thus attended by tensions and conflicts. Veteran behavior geneticists were eager about molecular technologies but not the superior attitudes they
perceived in the newcomers. Molecular geneticists thus reinforced the
archipelagic dynamics overall—though their high-handedness put off
human and animal researchers equally. The scientific tools, however,
didn't stick to particular islands. Molecular tools were used freely, and
as we will now see, just as much as they changed long-standing scientific
practices, the patterns of their use were also shaped by the field's forces.

Feet in Both Worlds

None of veteran behavior geneticists' misgivings about molecular geneticists should suggest they were unenthusiastic about the prospects of the
new methods. Field leader Robert Plomin wrote in 1990 that "the use
of molecular biology techniques will revolutionize behavioral genetics,"
and his cheerleading has continued to the present.[70] Veteran behavior
geneticists did not stand back and wait for molecular researchers to take
on behavior; they actively sought to bring it into their research. An im-

portant 2003 collection, *Behavior Genetics in the Postgenomic Era*, edited by Plomin, John DeFries, Ian Craig, and Peter McGuffin, showcased their work.[71] Building on a 2001 conference, the volume had fifty-four contributors and twenty-six chapters on research strategies, learning and memory in mice, cognitive abilities (intelligence, learning, reading, language), cognitive disabilities, psychopharmacology (including addiction), personality, and psychopathology. In this there was a tilt toward the psychiatric and medical side of behavior, and a clear effort to avoid more controversial topics like sexuality, violence and criminality, and race. This collection aimed to show how veterans and newcomers were working together bringing molecular research to classic topics.

Molecular research in behavior genetics focused on the identification of quantitative trait loci (QTLs). Quantitative traits vary continuously, like height or IQ, in contrast to Mendelian traits that have discrete forms, like Mendel's peas or genetic diseases like cystic fibrosis. Where Mendelian traits are usually due to the action of one gene, quantitative traits are believed to be caused by the contributions of many genes (plus environment). QTLs are places in the genome statistically associated with the trait of interest. Different methods and statistics (in both human and animal research) enable the search for QTLs; the hope is that identifying QTLs will let researchers estimate how many gene variants contribute to a trait, how big an effect the genes have, and, crucially, where in the genome the genes might be so their identity and function can be ascertained. Unlike quantitative genetics, which gives global estimates of genetic effects, QTLs are real pieces of DNA.

For skeptics the search for QTLs did not change behavior genetics, but instead was assimilated to the habits of scientific production that had long prevailed. As one animal behavior geneticist described the QTL research:

> It's a substitute for heritability, for a lot of people. For a lot of people at some point the goal of research was to estimate heritability of name it in, mice, humans. And then at some point it became QTL. And most of the QTL work as it has been done up till now, I find as scientifically sterile as the heritability research. . . . This only has value if you can get to the gene, if you can actually at some point say this is the gene doing it. . . . But nobody's going there. Look at what these people are doing, they do one QTL study after another, and say, "Oh, we narrowed [the location] down the end of [the chromosome] a little bit." . . . But again, my gut feeling is that these people don't care about

the gene. Before they had this figure for heritability, now they want to have a number for the chromosome. But they're not really interested in understanding what's happening.[72]

The complaint was that the identification of QTLs wasn't leading behavior geneticists to become interested in how the genes might function biologically to affect behavior. Instead, he charged, QTLs were being treated like heritability estimates; behavior geneticists were acting as if finding them is an end in itself. Racking up more and more of them appeared to become the practical goal.

These habits were not limited to QTL research, strictly defined. If QTL research aids the search for unknown genes, candidate gene studies proceed with a hypothesis that a known gene variant might be associated with a behavior. It would seem to point toward biological function. A recent review by political scientists Evan Charney and William English suggests otherwise.[73] They considered about six hundred papers on genetic associations with known variants in four genes (*MAOA*, *5-HTT*, *DRD2*, and *DRD4*) to 159 traits, more than one hundred of which were unambiguously behavioral. In a chart organized by trait, they list studies that have shown an association or a nonassociation with variants of each of the genes. First, their chart confirms the generally ambivalent record of the field: most positive associations are accompanied by nonassociations, and most associations without accompanying nonassociations are the only published study on the particular trait/variant relationship. But more crucially for the present point: this bird's-eye view reveals that researchers are casting about for any association they can publish. There are behavior genetics classics like intelligence and novelty seeking but also things like "choosing between greater reward or lesser punishment," "confirmation bias susceptibility," "consumer versus sales orientation," "cooperativeness," "creative dance performance," and "credit card debt," and that's just some of the "Cs"! Molecular genetics may have failed to meet expectations, but it has clearly succeeded in providing researchers with the means to crank out publications.

Another critic complained about the peripatetic inclinations of behavior geneticists:

It's been now ten years since the original gay gene study was published. And it's not been independently replicated by anybody else. . . . The person who was leading it no longer works on it. If it's supposedly true, we still have no idea about the developmental pathways. [laughs] So you can kind of ask, what purpose was served by engaging in that research if you're not going to see it through to some kind of meaningful end? And unfortunately a lot of the sort of single trait, single genetic trait, single genetic locus stuff has had the same fate. . . . [People] fire off papers and they don't stick with it, is the thing. It seems like a lot of the people engaging in the research don't actually have a commitment to understanding what really happens. [laughs][74]

This critic has been accused of opposing behavior genetics on political grounds, but how he articulated his objection was actually sociological. The problem was that behavior genetics, and science more generally, seemed to be rewarding the accumulation of superficial contributions rather than sustained focus on one topic.

These examples show that molecular genetics has been a protean resource for behavior genetics. It has driven renewed attention and interest and has invigorated the research agenda. Ironically, though, rather than transforming the field intellectually—ushering in a new era of mechanistic research, molecular genetics has been bent to the patterns of intellectual production in behavior genetics. In reference to a distinction from the last chapter, the aspirations of some to transform behavior genetics into a "cumulative" science that explains the link between genes and behavior mechanistically have foundered on the "accumulative" pattern of research that was energized by the new molecular methods.

Despite the different visions of a new era in behavior genetics, the move to molecularize research was always partial. Table 6.1 shows the numbers and proportion of articles published in *Behavior Genetics* by decade that listed at least one molecular genetics keyword in its abstract. On the one hand, the increase in articles engaging molecular genetics is clear: only one in twenty before the genomic era (post-1990s) and from a fifth to a quarter during it. When extending to articles that mention a keyword anywhere in the text (including references) the proportion rises, three-quarters to four-fifths. Most of the difference is due to keywords in the references, and this trend suggests that the vast majority engages the molecular genetics literature. But on the other hand, the table suggests that, even today, only about a quarter of articles are really *about* molecular genetics.[75]

TABLE 6.1. **Molecular genetic keywords in** *Behavior Genetics*

	Total articles	Number of articles with MG keywords in abstract	Percentage of articles with MG keywords in abstract	Percentage of articles with MG keywords anywhere
1970–79	378	23	6	8
1980–89	537	30	6	7
1990–99	596	111	19	31
2000–2009	679	174	26	73
2009–12	251	60	24	80

Note: This chart is derived from keyword searches of *Behavior Genetics* using the online database on September 11, 2012 (http://www.metapress.com/content/105485/). The keywords were: molecular, DNA, genomic, allele, linkage, QTL, SNP, transgenic, knockout. They were combined using the Boolean "or" operator. Articles with keywords in abstracts are likely to substantially engage molecular genetics, while keywords in full text are likely to pick up studies with passing mention or with molecular keywords in their references.

Many factors have contributed to the persistence of nonmolecular behavior genetics. We will see how practices of quantitative genetics were adapted to continue to enable the production of scientific capital even as the new technologies were ushering in changes to the structure of the field. Nonmolecular methods, and in particular the much maligned heritability statistic, acquired new life in relation to the new molecular methods. First, molecular researchers used heritability estimates to strategize about molecular genetics research. As a leading molecular geneticist said at a behavior genetics summer school I attended, heritability helps "to get your grant funded." Funders, he said, care about explaining variance in a disease or phenotype because they want to fund things that will have a large effect, that is, a large public health impact.[76] A behavior with a high heritability was assumed to be one where the genes would be easy to find and where, somewhere in the future, genetically informed interventions would be likely to be effective.

Second, as Arribas-Ayllon and colleagues have pointed out, since molecular genetics has run into trouble, researchers have repeatedly asserted heritability estimates for behavior to defend the rationale for genetic research.[77] They use past heritability findings to claim that "complexity" is the problem, and to direct attention away from the crumbling authority of the dominant paradigm of genetic causation. It is worth pointing out that critics have long contended that the connection between a heritability estimate and the probability of identifying molecular genetic markers is a folk notion and that conceptually and methodologically there is no connection.[78]

Third, and relatedly, heritability studies have persisted because they work. Behavior geneticists have an established infrastructure for producing them and publishing the results. As molecular geneticists struggle to identify reliable or robust molecular links to behaviors (or indeed most other traits), behavior geneticists continue to crank out studies based on quantitative genetics. In fact, while molecular researchers cite high heritabilities as a reason not to despair, these scores also represent a highly visible reminder of their failures. An entire literature has emerged to puzzle over "missing" or "hidden" heritability—the gap between the high double-digit estimates of genetic effects from family studies and the low single-digit molecular findings.[79] Indeed, it is ironic that Claire Haworth and Robert Plomin, the champion of molecular genetics, have recently trumpeted the "bright future" of quantitative genetics in the molecular era.[80]

Behavior Geneticists as Holistic Environmentalists

Even as behavior geneticists eagerly pursued molecular genetics research (and continued with heritability-based studies) the competition with molecular and psychiatric geneticists stoked anxiety. Would behavior geneticists be able to produce meaningful research with such high-powered competitors in the field? Would molecular research devalue quantitative genetics? In response veteran behavior geneticists began to reimagine their expertise as against the genetic reductionism and determinism of molecular genetics. And this filtered into their research as they reworked it to focus on measuring environmental effects and interactions rather than genetic determination.

An important component was to position themselves as more concerned with the phenomenon of behavior and the whole person than their molecular genetics counterparts. For example, one psychological behavior geneticist marked the following distinction with psychiatric researchers:

A clinician, a psychiatrist, being a medical doctor, they tend to just totally believe in the sanctity of diagnoses. So they don't have any problem with measurement. You just follow the DSM criteria, you tick the boxes, "Yes, they're schizophrenic." "No, they're not." And they believe these things; you know, where psychologists believe behavior is much more complex and they can't—

you know, a diagnosis is a very rough way of getting at this. . . . So [psychiatrists] weren't plagued by any doubts there. And doctors aren't much plagued by doubts, anyway.[81]

The speaker's argument was that the behavior geneticists, coming from psychology, are much more sensitive to the complexity of behavioral traits than psychiatrists (and by extension molecular geneticists). This sensitivity, he claims, has led them to be more epistemologically reflexive. Behavior geneticist Thomas Bouchard put this sentiment more boldly when he said in an interview: "Molecular genetics looks at genes, not whole, live human beings. Twin studies add a very necessary human element to genetics and that needs to be recognized."[82] An even more radical position has been advanced by Eric Turkheimer, who has called on behavior geneticists to drop the charade that biological reduction is the aim of their work, to admit, simply, that they are social scientists, and to accept that complexity and ambiguity will always accompany their work.[83]

What should be noted here is that behavior geneticists positioned themselves in precisely the opposite terms than they used in their competition with psychologists and social scientists before the molecular era. Then they saw it as a point of pride to reduce behavior to genetic causes; they dismissed careful study of the complexity and dynamics of behavior and the person. With these positions, psychologists often perceived them as crude and naive—perceptions they came to have of molecular scientists.

Instead of antagonizing molecular geneticists, this repositioning sometimes allowed behavioral geneticists to cooperate with them more effectively. Here is an example. According to several of my interviewees the UK Medical Research Council's support of molecular genetics research in psychiatry and behavior was controversial in the mid-1990s because of anxieties among the public and critical scientists about genetic reductionism.[84] One psychological behavior geneticist said that this controversy led to the hiring of behavior geneticists at a UK research center:

At a very political level, it was a hire that would counteract the public perception—or the MRC had a concern that there was a public perception that they had gone too far, too fast in the area of molecular genetics. So they needed to be able to show that there was a counterpoint to that within this center. . . . The reason we were hired here is because the MRC, after funding this sector, got cold feet about its public reputation going too far in the molecular genetics direction.[85]

In this climate, behavior geneticists who took behavioral measurement and environment relatively seriously could help embattled molecular geneticists soften their image. Such a perception of mutual benefit likely also helped facilitate interdisciplinary conferences and research. We can see in this another dimension of "giving the field away." This "environmental" repositioning went beyond image and had an effect on behavior geneticists' research. This same speaker put the point clearly:

> I think that traditional behavior geneticists who had invested a lot of their careers and a lot of money in building twin samples, and who now have these twin samples as their major research tool, were kind of panicked by that belief that appeared with the announcement of the Human Genome [Project conclusion] that their research was worthless. And so what I see is an increasing interest among that community of scientists who have big twin studies and adoption studies . . . using them as tools to look at environmental effects."[86]

Behavior geneticists' competition with molecular researchers produced anxiety about the devaluing of the resources and expertise they'd accumulated. This made a new focus on the environment expedient.

Behavior geneticists had long claimed that environmental effects were an important part of their research; after all, so the argument went, if most traits are about 50 percent heritable, that means the other 50 percent of the variance is due to environmental variance. From claims like this (mobilized mainly, it seemed, to defend against charges of genetic determinism), behavior geneticists focused closely on partitioning environmental variance into the shared and nonshared varieties. This was the period, discussed last chapter, where behavior geneticists pushed the interpretation that variations in normal parenting and other forms of socialization don't matter much. Studying the environment typically meant further and further constraining the space for traditional notions of environmental causation.

But since about 2000, as the competition intensified with molecular genetics (and chinks in its edifice became more apparent), behavior geneticists have taken environmental study more seriously. For example, psychological behavior geneticist Eric Turkheimer and colleagues conducted an important study about the dependence of IQ heritability on socioeconomic status (SES).[87] For high SES kids they found, like many behavior geneticists before them, that IQ variance is overwhelmingly due to genetic variance. What was new was that for low SES kids environmental variance explained nearly all of the IQ variation. Turkheimer interpreted this as showing that poor kids' environments aren't stable enough for genes to matter, which argues against the view that their environments are "good enough" to let kids reach their potential.

Another example came from a psychological behavior geneticist who described several environmental studies about antisocial behavior. One addressed the hypothesis that having a household with an absent father leads kids to be antisocial: "when we controlled for genetic transmission, what we found is that there is no association between father absence and children's antisocial behavior. So it appears that that's just an artifact of men with an antisocial predisposition being more willing to abandon their kids."[88] The behavior genetic model allowed the testing of an environmental hypothesis. Finally, behavior geneticists Diane Dick and Richard Rose led a team on a series of studies on smoking and drinking behavior that compared twins and matched nontwin controls in a variety of school and neighborhood settings. In one study they found that neighborhoods and schools greatly affected these behaviors and in another that whether or not genetic differences affect behavioral variation depends greatly on qualities of the environment (in settings with easy access to alcohol or lots of peer role models genetic differences influence behavioral differences more).[89]

Behavior geneticists, not molecular geneticists, were also the leaders in the molecular research on gene/environment interactions. The team led by Avshalom Caspi and Terrie Moffitt produced several of the most celebrated. I mentioned one earlier: the finding that the degree to which stressful experiences translate into depression depends on the form of a particular gene. Another study asked why child abuse could lead to different degrees of antisocial behavior. They found that mistreated children with the "low activity" form of the monoamine oxidase A gene exhibit much more antisocial behavior than mistreated children with the "high activity" form.[90] Behavior geneticists were attracted to this research in part because they were historically sensitive to the charges that their research promoted genetic determinism. They believed such studies showed that their claim to be interactionists concerned with nature *and* nurture was more than just talk. Further, in both of these studies the effect of genes alone wasn't significant; genes only mattered in con-

cert with particular experiences. Caspi and Moffitt hinted that this in-
terdependence, seldom measured, might be a reason why molecular genetic studies had been disappointing.[91] Implicitly, this was an argument
that social-science-oriented behavior geneticists, because of their better
grasp on environmental measurement, would beat molecular geneticists
at the gene-hunting game.

Sociologist Thomas Gieryn analyzed how John Tyndall, superinten-
dent at the Royal Institution in Victorian England, sought to define science publicly in contrast to religion and engineering: "Scientific knowledge is empirical when contrasted with the metaphysical knowledge
of religion, but theoretical when contrasted with the common-sense,
hands-on observations of mechanicians; science is justified by its practical utility when compared to the merely poetic contributions of religion,
but science is justified by its nobler uses as a means of 'pure' culture and
discipline when compared to engineering."[92] Behavior geneticists similarly described themselves in seemingly contradictory terms depending
on the context. Battling psychologists, they sold their science in terms
of reductionism and biological determinism, but in distinguishing themselves from molecular geneticists they emphasized holism and environmentalism. Tyndall's efforts, according to Gieryn, were almost wholly
ideological, but behavior geneticists altered their practices during these
different efforts at self-definition. What's interesting is that although the
main research paradigm remained tightly focused on partitioning variance into genetic and environmental sources, that narrow range of practices could be conceived in almost opposite terms.[93]

Conclusion

The incorporation of molecular genetics into behavior genetics was thick
with ironies. Molecular genetics was supposed to launch a new era, to
revolutionize behavior genetics. Specifically, it was supposed to move the
field beyond quantitative genetics and heritability to the action of real
genes. With molecular genetics failing to deliver, not only has heritability persisted, but molecular researchers grasp onto it as a lifeline to justify their work. The molecular future used to stand as a rebuke to heritability, now the steady way heritability delivers results has become a
rebuke to roller-coaster inconsistency of molecular research. Molecular
genetics was also supposed to revolutionize the style of behavior genet-

ics research: attention would become focused on mechanistic causes of behavior. But instead, molecular research adapted the peripatetic habits prevailing in behavior genetics, leaping from method to method and behavior to behavior.

Another promise of molecular genetics was that it would integrate behavior genetics and unify it around a new consensual paradigm. Further, an infusion of new experts would clean up the shop, since many scientists viewed behavior geneticists as "mere" psychologists trying to elevate themselves with tools they weren't competent to use. But the molecular era has been characterized by the proliferation of controversies and the reinforcement of old professional and conceptual divides. New entrants didn't mesh with the veterans, and while exchanges among islands were vigorous, the archipelago persisted. In addition, far from bringing caution and reserve to the discourse around behavior genetics, the new experts often spoke with the greatest recklessness and intoxication (a theme I pick up in the next chapter). Even the failures of molecular genetics were shot through with irony. Responses to disappointed expectations have generally entailed the investment of greater resources—more money, larger samples, bigger collaborations, new levels of technology. These disappointments may have been a blow to the field's cognitive authority, but they have produced a boom in practical activity and investment.

Overall, then, the advent of molecular genetics has turned out to be more reproductive than revolutionary for behavior genetics. Prevailing structures of practice and social organization steered the incorporation and reactions to the technologies much more than being changed by them. Perhaps things would have been different if molecular genetics hadn't foundered. Different scientific opportunities would have certainly shifted participation and practices. But it isn't like the failure of molecular genetics hobbled active projects to transform the field socially or practically; technological problems were not the reason molecular enthusiasts neglected, for example, to try to integrate the scientific societies or to change styles of scientific production.

The story here complicates the notion that scientists' intellectual commitments are determined by the experimental paradigm or technological tools they employ. These matter, certainly, but their effects are mediated by field-level structures of distinction and recognition. For example, when behavior geneticists were struggling for recognition against psychologists and social scientists, they positioned themselves as hard reductionists, genetic determinists (though a circumspect variety), and rigorous quantifiers of behavior. But when competing with molecular and psychiatric geneticists, they touted their concern with the environment and interactionism, the complexity of behavior, and the integrity of the whole person. More than just talk, these shifts underwrote changes in their research. What is more, the disappointments of the molecular genetic technologies have generally led researchers to become more, not less, invested in them. And it was the habits and rewards of the field that led them to be deployed in an "accumulative" mode rather than to integrate research practice in other ways. It is the interaction of the technologies and paradigms with the dynamics of the field that produce intellectual commitments; neither is sufficient on its own.

Finally, it is worth reflecting briefly on the unintended scientific consequences of molecular genetics. Anthropologist Stefan Helmreich has written about evolutionary biologists' efforts to incorporate genomics in order to make their taxonomical schemes more unified and rigorous and to resolve previously obscure evolutionary relationships among life forms.[94] Instead of securing taxonomy, genomics undermined the very idea of taxonomy and threw the concept of evolutionary relatedness into doubt; rather than a stately tree, evolutionary relationships (especially of microorganisms) have become a rhizomatic tangle. The failures of molecular genetics in behavior genetics may have similar unintended consequences and transform the way we think about behavior, genes, and environment. But it may just undermine the imperial ambitions of genetics. As one skeptical behavior geneticist put it: "The impact of genetics on the overall undertaking of behavioral science has been and will be less than everybody thought it was going to be. And that, in fact, what happens when the traditional problems in psychology meet a new technology, it's not that the technology transforms the old problems into new science, it's that the complexity of the problems transforms the technology into social science of the old kind."[95] Perhaps genomics will be more affected by its encounter with behavioral science than the reverse. Surely, this is but one more prediction, albeit one that isn't tacking upwind.

Responsibility, Notoriety, and Geneticization

Behavior genetics has always had enormous public appeal. Sandra Scarr once bragged that in contrast to behavior genetics' often jaundiced reception by scientists, "there has always been a reservoir of sympathy in the public consciousness for what we have been promoting."[1] In their classic *The DNA Mystique*, sociologist Dorothy Nelkin and historian Susan Lindee documented hundreds of examples of the ways that ideas from behavior genetics have inspired pop cultural portrayals of human character and capacities, social differences, and notions of responsibility.[2] Although the trend has quieted a bit, through the mid-2000s, nearly weekly headlines would trumpet a new genetic discovery, as often as not about some behavioral trait. There have been several dozen popular science books—both critical and celebratory—published on the topic since 1970.

The incredible degree of public attention to behavior genetics has been a bone of contention. Critics have complained about the dangers of "geneticization": the tendency to understand human behaviors and differences as genetically determined and resistant to change.[3] Sociologist Barry Barnes and philosopher John Dupré have suggested the succinct label "astrological genetics" for a broad range of discontent: the idea that "our entire moral nature, and the fate it brings upon us, may be attributed entirely to our inheritance"; that there are "'genes for' politically sensitive traits, like intelligence, criminality, homosexuality, addiction and substance abuse, and propensity to violence"; and that genetics is "the properly scientific way of explaining such traits."[4] Critics believe geneticization and astrological talk dangerously reduce human nature and experience, misleadingly imply biology explains everything, and naturalize human problems, which either justifies inaction or suggests techno-medical solutions.

For their part, behavior geneticists are also dissatisfied with the public portrayal of their science, but for less focused and more varied reasons. They have long complained that their work is too frequently linked to the specter of eugenics and other dystopian possibilities; there's too much focus on issues of race that they see as scientifically marginal; and their critics, often through journalistic norms of balance, are given too much attention. They too worry about aspects of geneticization. As one psychological behavior geneticist put it:

> People in society, people in fairly important places like judges, policy makers and so forth want very simple explanations of things. So if there was a heritability of something they would interpret that as, you know, we shouldn't have Head Start programs, we shouldn't have rehabilitation programs. It's genetic, there's nothing you can do about it. I felt like, this doesn't have anything to do with biology . . . it's really about culture. It's about what we're willing to understand and what we learn and how we're willing to see things.[5]

People believe all sorts of inaccurate things about genetics, but behavior geneticists don't want to be blamed for cultural tendencies toward ignorance and exaggeration.

Previous chapters have focused on conceptual disputes. Conflicts about the communication of those ideas to the public are the topic of this one: What are behavior geneticists' responsibilities regarding the communication of their work? Do they have an obligation to correct each other's representations? How should they distinguish between legitimate and problematically provocative ways of talking about behavior genetics?

Critics usually accuse behavior geneticists of two types of irresponsibility. First, they say, too many behavior geneticists exaggerate and oversimplify their claims, both to other scientists, but especially to public audiences. Second, behavior geneticists don't do enough to correct misperceptions or challenge incorrect claims on the field's behalf. If, for example, most of them believe race researchers' claims about genetic differences cannot be substantiated, they should publicize that view.

Behavior geneticists, not surprisingly, typically resent these accusations. They see the exaggerations and oversimplifications as largely not their fault. Journalists misunderstand, oversimplify, misquote, and sensationalize their work (a common complaint among scientists). Behavior genetics ideas are also inherently complex, and understanding them depends on biological and statistical knowledge that few journalists or readers have. Behavior geneticists see themselves as expending a lot of energy on efforts at explanation, only to be criticized. Most would admit, finally, that excitement about the science has led sometimes to "evangelism," and a small minority has always sought to be offensively provocative. But much of the overexuberance is a thing of the past, and the dedicated provocateurs aren't going to listen to reason anyway.

These two views of responsibility among behavior geneticists, while opposed, share a common framing. The first view says essentially that behavior geneticists simply have to try harder at being responsible. The second says that trying harder won't matter because the problem is outside the field, in the media and broader society. The focus of my argument, and what these opposing views neglect, is that the field's history and structure mediate "responsibility" by shaping its definition, possible communicative practices, and relationships to the media and public. In short, provocative public communication has become a source of symbolic capital written into the field's logic. It is encouraged by prevailing definitions and practices of responsibility, the field's structured anomie, and an eager cultural and media marketplace.

This chapter shows the developmental and structural logics that have shaped the conceptualization and practices of "responsibility" in behavior genetics. As the companion to chapter 6, this chapter focuses on roughly the same time period, 1990 to the present, though it connects to crucial earlier moments. The first part shows how the experience of controversy encouraged behavior geneticists to identify "responsibility," not as social responsibility, but as a responsibility to science, to defend scientific freedom. Public controversies in the 1990s inculcated a sense of obligation to participate in "social implications" projects convened by bioethicists. These projects, I show, failed because they couldn't help but reproduce the conflicts they intended to overcome.

If such institutionalized projects falter, then informal efforts face even more challenges. The field's fragmented and anomic structure makes it inhospitable to policing, critique, and even serious communication. Such critical communication has been abandoned by most scientists and as a result has been effectively monopolized by the field's cadre of longtime critics. The chaotic and controversial character of behavior genetics has made it poorly situated to allow scientists to pursue recognition and authority in traditional terms. But the upside for many scientists is that it does allow them to speak directly and confrontationally to matters of intense public curiosity. I call this the symbolic capital of notoriety—in many ways it is the opposite of responsibility, and I show how it has become embedded in the field's logic.

The final section returns to the matter of behavior genetics' broader social and cultural impact through geneticization. I argue, on the one hand, that the field's limited authority makes its impact weaker than commentators have often feared. On the other, these effects can still be very real, varied, and surprising, stretching from individual's self-conceptions and performance to the cultural logic of neoliberal politics.

Scientific Responsibility or Intellectual Freedom?

A perennial ethical debate in science concerns the relationship between intellectual freedom and responsibility. The freedom to study any topic no matter how scandalous or blasphemous is a cornerstone of the scientific ethos. Galileo and Darwin are heroes in the mythology of science as much for their challenges to religious authority as for their ideas. But ideas of responsibility have been strong as well. Shouldn't science serve human improvement? Should purely destructive inquiry be pursued just because it can be? Darwin famously fretted about the radical implications of his work and long delayed publication. This debate has run in behavior genetics, too; its internal debate, churned by controversies, has developed in relation to those of other communities. We will see how behavior geneticists have come to define responsibility in terms of an absolutist ethos of intellectual freedom.

In some ways behavior genetics owes its very existence to a strong commitment to scientific responsibility. As I showed in chapter 2, behavior geneticists considered questions of racial differences and eugenics to be outside the field's intellectual purview for both political and scientific reasons. They understood that racists would seize upon any research on the topic, but they also felt the difficulties of separating genetic and environmental influences in a racially stratified society were too scientifically problematic to warrant attention. They understood that too much freedom—an open invitation to study race and eugenics—would have sunk the young field.

With the Jensen controversy, under intense pressure from all directions, behavior geneticists shifted from this combination of freedom with responsibility to a more absolutist interpretation of intellectual freedom. Responding to the controversy in 1972, educational psychologist Ellis B. Page organized fifty prominent behavior geneticists and allies in genetics and psychology to sign a statement published in *American Psychologist* denouncing "suppression, censure, punishment, and defamation . . . against scientists who emphasize the role of heredity in human behavior."[6] Absent among signatories were the architects of behavior genetics' boundary against race research, most notably Theodosius Dobzhansky, the field's scientific and moral giant. Dobzhansky had argued that the statement would be stronger if it included scientific responsibility: the scientist's "duty to state publicly that the misuses of his research are just that—misuses."[7] This request was ignored and Dobzhansky did not sign. Dobzhansky's ideas about responsibility failed to find a foothold in behavior genetics. Calls for responsibility by Jerry Hirsch and other internal critics got lost among the attacks on heritability. And other behavior geneticists, seeking to preserve as much space as possible for their besieged field, may have worried that endorsing such responsibility could have added further unwelcome constraints.

Many prominent behavior geneticists vigorously embodied the anti-authoritarian cultural ethos of the 1960s. Sandra Scarr emphasized her membership in the American Civil Liberties Union, "to assure [herself] and others that an interest in genetic differences did not necessarily go with antidemocratic politics."[8] Thomas Bouchard was touted as a leader of Berkeley's free speech movement.[9] And psychological behavior geneticist David Lykken wrote in his memoir of his successful lawsuit against the police of his hometown near Minneapolis for a raid they conducted on his house during an anti–Vietnam War gathering.[10] Far from the tweedy apologists for an elite-controlled social order as they were often accused, behavior geneticists viewed themselves as alienated from authoritative institutions. These included disciplinary authorities, whom they believed imposed an environmentalist orthodoxy; university authorities, whom they saw as more interested in smoothing political conflict with egalitarian platitudes than defending intellectual freedom; and activists, who, in their view, fought to institute dogmatic views, not individual free thinking.

Behavior geneticists were advancing one particular version of the 1960s ethos. In *Disrupting Science*, sociologist Kelly Moore mapped the sources and evolution of scientists' postwar repertoires for political action.[11] In the late 1960s and 1970s, members of the Science for the People (SftP) network were the most visible scientist activists. They were politically radical: sensitive to science's role in erecting and maintaining economic and political structures that promoted warfare, violence, exploitation, and inequality. They defended an ethos of social justice and aimed to make science serve "the people." Behavior geneticists embodied a very different spirit, and not just because some of their critics, like geneticists Richard Lewontin and Jonathan Beckwith, were SftP members. Behavior geneticists promoted 1960s values of freedom and individuality but mostly ignored social justice. Although they drew from the revolutionary rhetoric of the day, their "activism"—a label they would not embrace—drew from repertoires Moore calls moral individualism and institutional reform, which SftP rejected. Behavior geneticists wanted disciplines to respect their work and universities to defend their rights, and as individuals they would stand as stalwart champions of those ideas and rights.

These views on intellectual freedom and responsibility continued to structure behavior geneticists' responses to controversy. Many saw it important to bear witness in a group letter to the *Wall Street Journal* to the scientific value of *The Bell Curve* despite its political unpopularity.[12] Those who felt *The Bell Curve* was irresponsible in its presentation of the science or its political interpretations declined signing the letter but did not voice their views publicly.[13] Within the field only the small cadre of established internal critics like Jerry Hirsch and colleagues spoke up.[14] With regard to J. Philippe Rushton's book *Race, Evolution, and Behavior*, one behavior geneticist told me that even though he and others didn't believe Rushton's ideas could be substantiated scientifically, "I think, the field has strongly defended his right to undertake his scholarship, and I think that's appropriate. I think it's really appropriate."[15]

When Glayde Whitney made a racial polemic as his BGA presidential address in 1995, there was debate about what the society should do. Intellectual freedom absolutists felt Whitney could say what he pleased, and the BGA had no right to censure. Others felt that speaking in his capacity as BGA president, Whitney's views would be taken as official or representative. This made it a collective matter, no longer one of individual intellectual freedom. According to one interviewee, the discussion—occurring via e-mail and online bulletin board—quickly became acrimonious.[16] The intellectual freedom position won, and there was bad blood on both sides. Several members resigned from the BGA, including the incoming president, but Whitney remained unsanctioned. The executive committee passed a motion saying the president's views are not official policy of the BGA and that "members are not encouraged to express their personal political and moral views," but these statements were not publicized.[17]

Efforts by some behavior geneticists to assert a middle-ground version of responsibility have foundered. Intelligence researchers John Loehlin and Earl Hunt have suggested that behavior geneticists not only use the most rigorous methods possible but also communicate their findings with caution and precision especially on socially sensitive topics.[18] Critics on the left, like geneticist Jonathan Beckwith, see this as inadequate; to them, scientists have a collective responsibility to educate themselves about and speak against the misrepresentation and misuse of science.[19] On the right, race researcher Linda Gottfredson argues that "scientific responsibility" puts a double standard on controversial topics, and the assertion of collective intellectual standards (let alone moral standards) is the road to censorship.[20] Caught in this pincer, it is no surprise that an absolutist view of intellectual freedom has become dominant among behavior geneticists.

Publicly, behavior geneticists often claim responsibility as a motivation—helping people is a goal. And ideas of public responsibility, like those articulated above, guide others privately.[21] But behavior geneticists understand their primary responsibility as defending the prerogatives of science against an often-hostile world. Further, iconoclasm is written deeply into behavior geneticists' public image.[22] When the question of responsibility is called during moments of controversy, the freedom position has come to prevail.

Scientific Responsibility Institutionalized

Thus behavior geneticists have refused to compromise their intellectual freedom with concessions to public responsibility. But as the early 1990s brought the rapid-fire announcement of new behavior-linked genes, the "gay gene" most prominently, as well as the *Bell Curve*–centered race controversies, they came to take more seriously the notion that some positive expression of scientific responsibility was required of them. The supposed social implications of their work were a major selling point, but these had a tendency to produce public controversies that boiled over into full-blown scandals that threatened the field's legitimacy. From the mid-1990s to mid-2000s scientific responsibility became defined as *participation* in ethical deliberation projects.

In 1997 the American Society of Human Genetics (ASHG) assembled a group of psychological, psychiatric, and molecular geneticists to develop a statement about past accomplishments and future directions in human behavioral genetics. Much of the statement was devoted to reviewing the classical twin and family methods and the emerging molecular ones. Interestingly, with the wounds of the *Bell Curve* controversy still raw, the ASHG statement discussed "group differences" but avoided any mention of race, choosing instead to discuss a hypothetical example of gender differences in "emotional stability." The statement ended by claiming geneticists have an "obligation to participate in educating the public, in nontechnical language, about the complexity of human traits and the simple facts of human variation."[23]

When the Human Genome Project was first funded in 1990, its leaders made the unprecedented move of establishing a sizable budget for the consideration of ethical, legal, and social issues (ELSI). High on the list for the emerging core group of ELSI scholars was to consider the implications of behavior genetics. As such, they began a series of projects that sought to bring together behavior geneticists, ELSI scholars, and sometimes the field's critics to foster different kinds of public understanding and conversation (see table 7.1). Behavior geneticists were eager to demonstrate their responsibility by participating. As one molecular geneticist told me, "there's always going to be a politics around [behavior genetics and also racial differences], and I think it's vital for us to continue to discuss it in public policy forums. That's why I take part in those things."[24]

These projects took a few closely related forms. All were invitational events that brought together behavior geneticists with scientific critics and social commentators. Those at Maryland and Cold Spring Harbor were academic conferences, single events occurring over several days. The intersection of scientific problems and social implications were their focus, but communicating to the lay public was not their priority. The PBS project was a single, televised event that focused entirely on public communication through a moderated panel discussion format. The Nuffield and Hastings projects were sustained efforts that gathered groups of scholars for series of meetings over several years seeking to engage

TABLE 7.1. **Events promoting the public understanding of behavior genetics**

Event title	Sponsors	Dates	References
Research on Genetics and Criminal Behavior: Scientific Issues, Social and Political Implications	• School of Public Affairs, University of Maryland	1995	Wasserman (1996)
The Genetics of Human Behavior	• Cold Spring Harbor Laboratory, NY	1995	
Genetics and Human Behaviour: The Ethical Context	• Nuffield Council on Bioethics (UK)	1999–2002	Nuffield Council on Bioethics (2002)
Our Genes/Our Choices: Genes on Trial	• Fred Friendly Seminars • Public Broadcasting Service (PBS)	2002	Fred Friendly Seminars, Inc. (2002)
Crafting Tools for a Public Conversation about Behavioral Genetics	• Hastings Center for Bioethics • American Association for the Advancement of Science • Human Genome Project (ELSI)	2000–2002	Baker (2004); Parens (2004); Parens, Chapman, and Press (2006)

participants in an ongoing dialogue and to culminate in a series of publications accessible to nonspecialist audiences.

One of the positive, and seemingly unexpected, outcomes of these events was to mitigate some of the mistrust between behavior geneticists and their scientific critics. A critic who helped organize the Cold Spring Harbor event said, "I do think things like that are really important. And I'd like to think that as a result of the interaction, I mean, some of the things I was worried about on the part of people doing the work, I was allayed on as a result of interacting with them. I'd like to think the interaction goes the other way too."[25] The face-to-face interaction helped him understand the motives and concerns of people he had otherwise only encountered in print. A psychological behavior geneticist pointed to similar benefits: Richard Lewontin, one of behavior geneticists' most intransigent enemies, "turns out to be more reasonable [than one would think]. He has when I have spoken to him."[26] Such positive possibilities reinforce again the problems of fragmentation in behavior genetics, how the lack of opportunities for interaction fosters adversarial relationships.

Unfortunately, this way of organizing scientific responsibility stimulated far more conflict than it quelled. The Maryland conference was born under a cloud. Fredrick Goodwin, the NIMH director whose advocacy of research on crime and biology had stimulated the conference, compared the problem of urban crime to the disorder of the "jungle." Members of the Black Congressional Caucus were offended at what they took to be a racist description of blacks as criminal monkeys, and funding for the conference was delayed for three years.[27] When it finally occurred, protestors disrupted the event, denouncing the biology/crime link as racist and an invitation to eugenics.[28] This event was attempting to avoid scandalous science and communication, yet it was interpreted in precisely those terms.

In their very structure, these events tended to build in the conflicts that they were designed to overcome. Each was structured to lay out the scientific "basics" first and then to explore the implications. There was a division of labor in participation: behavior geneticists, scientific critics, and specialists in "implications." One effect was to highlight the separation between what C. P. Snow called "the two cultures." Philosopher Erik Parens, leader of the Hastings Center/AAAS project, wrote,

> Those of us from the humanities were sometimes intimidated by the languages of statistics and genetics. Sometimes the behavioral geneticists were frustrated that those of us in the humanities and social sciences did not take more time to learn the science. The behavioral geneticists also seemed (to this "humanist," anyway) impatient with the languages and concerns of the scholars in the humanities and social sciences.[29]

Critic Jonathan Beckwith, normally sympathetic with humanists' views, lamented their "mistakes in their interpretations of genetics research, making it easier for the geneticists to dismiss their suggestions."[30] A not insubstantial level of scientific competence became a de facto condition for speaking about behavior genetics, but humanists had trouble asserting authoritatively that analyzing "implications" demanded equivalent expertise.

The events couldn't help but frame the problem as science *versus* ethics and implications. This fit with behavior geneticists' views of the situation: their science produced a set of facts. The problem was how nonscientists appropriated them. "Weigh against a strong interpretation of the data, not against the data itself," said one psychological behavior geneticist.[31] Beyond explaining their science's meaning as clearly as possible—the rationale for their participation—behavior geneticists wondered why they should be held responsible for social misuses. During these events critical participants sometimes tried to break this frame to argue that "implications" also infuse scientific choices of, for example, behaviors to study, the very goal of separating genetic and environmental effects, and the lack of attention to behavior-modifying interventions. Raising these questions tended to look like an attack on the motives of behavior geneticists, and often by those who it seemed couldn't get the science "right."[32] Thus these efforts were caught in a dilemma: either accept the separation of science and implications or reject it and risk mutual antagonism.

The controversial character of behavior genetics and the adversarial structure of the projects' membership led leaders to seek points of agreement and capture the middle ground.[33] The Nuffield Council's report made this attempt.[34] The behavior genetics journal *Genes, Brain and Behavior* hosted a review symposium. The first reviewer, David Hay, a psychological behavior geneticist, thought the report too dependent on the views of the field's opponents, objected to the suggestion that the field be held accountable for completely hypothetical misuses of its findings, and worried that it would lead to funding and institutional review problems for the field.[35] Doug Wahlsten, an animal behavior geneticist and longtime internal critic, thought the report too sanguine about the field's research, especially the meaningfulness of heritability estimation.[36] The middle-ground compromise that the report sought turned out not to satisfy anyone.

The Hastings Center/AAAS project fared worse, basically collapsing according to several participants.[37] Although final publications were salvaged, most of the behavior geneticists withdrew when they became frustrated with what they considered an unfair focus on heritability controversies and criticism of their science. Given these circumstances this outcome is not very surprising. We can imagine behavior geneticists' thinking: What other field constantly has to appear before tribunals of its enemies to explain itself and justify its existence? What other field constantly has to be reminded of the follies of its ancestors that have nothing to do with its current goals and aims? In what other field do members make goodwill efforts to address their detractors and educate the public, only to have their good intentions backfire time and again?

This way of organizing scientific responsibility in behavior genetics seems to have been judged a failure. Contradictions undermined participation, controversy was more likely to be amplified than agreement reached, and few seemed satisfied with the results. Moreover, these projects occurred at a particular moment in time, and none like them have been attempted in the last decade. The story here parallels what anthropologists Paul Rabinow and Gaymon Bennett found in their work with synthetic biologists to incorporate ethical deliberation into the research process.[38] Scientists had a basic willingness to cooperate with the effort but were unwilling to collaborate. Synthetic biologists, like behavior geneticists, were unwilling to break the framework that separates "science" from "implications." The difficulties of institutionalizing scientific responsibility are not, therefore, specific to behavior genetics. Yet the failure is biting because the forces of anomie, to which we now turn, undermine the possibility of alternatives.

Responsibility in an Archipelago

Scientific responsibility is usually discussed in terms of individual commitment, but far more important in behavior genetics was the erosion of social structures that make responsible commitments and practices possible. In the 1990s and after, behavior genetics had become thoroughly fragmented at several scales. Yes, a core mainly comprised of psychologists, psychiatrists, and animal researchers still held onto "behavior genetics" as an identity and a label, but there were growing numbers of researchers doing science or theorizing in ways that the world would recognize as "behavior genetics" without any genuine engagement with that core group. Few social barriers or disciplinary structures in this heterogeneous space regulated the entry or conduct of would-be participants. And there was no sense that an authority had to be respected if someone was inclined to speak on behalf of the field. While these possibilities for provocation were increasing, fewer scientists (inside or outside behavior genetics) engaged in policing practices. These became effectively "monopolized" by a small cadre. In short, behavior genetics lacked structures to inculcate scientists' responsibility and afforded few tools to exercise control should it be desired. It is worth tracking how this disorderly situation played out on different levels.

First, leading geneticists have been eager to promote and pontificate about behavior genetics, but few have been moved to invest their energy or resources in actual research. As I showed in the previous chapter, field leaders, such as James Watson, Walter Gilbert, and biochemist and *Science* editor Daniel Koshland, used the dream of behavior genetics—unlocking the secrets of human nature and solving social ills like crime, homelessness, and madness—to justify massive government investments in the Human Genome Project. Working in a promotional mode and unmoored from the presentation of actual research, they never dealt with the criticisms and difficulties of these ideas. Promoters tended to speak in an "astrological" mode: genes are the human essence and they set the fate of behavior and disease. For them, behavior genetics was largely a symbolic resource. It was hardly a real scientific endeavor one might pursue. It could not command care in their representations, much less their participation when actual controversies emerged necessitating public interpretation.

At a second level, several scientists have done controversial work on behavior genetics though have had little engagement with behavior geneticists. Dean Hamer embarked on his "gay gene" study without interacting with behavior geneticists, claiming that little of their research was relevant to his.[39] Similarly, when Bruce Lahn suggested that average differences in intelligence between blacks and whites might be due to mutations in genes linked to brain development that emerged after ancestral humans migrated out of Africa, he made no reference to the literature on these contentious ideas.[40] In these cases research to replicate the claims was done—and confirmation failed for both—but scientists in the field were reticent to offer overall assessments for the public.[41] Historian Sarah Richardson has argued that this public silence was especially noteworthy regarding Lahn's work because it clearly failed to meet geneticists' own standards for making evolutionary genetics claims.[42] These and other scientists have spoken on behalf of behavior genetics with minimal social or substantive relationships to the field.

In addition, much of the public conversation about behavior genetics has occurred in the realm of popular science books, a domain to which most behavior geneticists have very little access. There have been dozens of books for general audiences published about behavior genetics. Among those by advocates are *Nature via Nurture* by journalist Matt Ridley, *Born That Way* and *The Nurture Assumption* by writers William Wright and Judith Harris respectively, and *The Blank Slate* by psychologist and public intellectual Steven Pinker.[43] Some of the most trenchant critiques have been published in books: Kamin's *Science and Politics of IQ*; Lewontin, Rose, and Kamin's *Not in Our Genes*; Hubbard and Wald's *Exploding the Gene Myth*; and Gould's *Mismeasure of Man*, to mention some of the most prominent.[44] In her analysis of evolutionary psychology in the UK media, sociologist Angela Cassidy has shown that popular books have been an important venue for the initial articulation (not just the later popularization) of researchers' basic scientific ideas.[45] In behavior genetics the pattern has been very different. When basic scientific ideas about behavior genetics have appeared in popular books, the authors have either been the field's critics or marginal members with a particular interest in provocation, for example, Rushton's *Race, Evolution, and Behavior* and Herrnstein's *IQ in the Meritocracy, The Bell Curve* (with Murray), and *Crime and Human Nature* (with James Q. Wilson).[46] Major books by "mainstream" behavior geneticists have been published by academic presses or stayed in the mode of popularization, like David Rowe's *Limits of Parental Influence* or Michael Rutter's *Genes and Behavior.*[47] The archipelago has extended into this domain of communication where the most inflammatory behavior geneticists are most prominent, others participate mostly by proxy, and the dominant style of presentation is both promotional and adversarial.

Within this multiply fragmented domain, behavior geneticists believed themselves to have very few tools for controlling vexing claims with which they disagreed. When asked what could be done about provocateurs, one animal behavior geneticist simply responded, "Don't ask him to lecture. Don't attend his lectures."[48] Another expressed annoyance at the unfairness that a few people who "wear black hats" are taken as representative of all behavior genetics:

> [A "black hat"] would never be taken as a spokesman by the rest of us. . . . But I can't spend my time and energy going out and interviewing with newspaper people for example and putting up my own web page saying, "Get these guys out of it; they don't represent us." That again, that attracts too much attention to them and it will make you talk about freedom of speech, the academic freedom. One of the prices that I'm willing to pay for academic freedom is to have . . . outliers.[49]

Statements like these demonstrate behavior geneticists' commitments to intellectual freedom, their long-standing disinclinations to police, and also the very real dilemmas of criticism that also may publicize and aggrandize those who are already savvy scandal makers. (We should recall, however, that behavior geneticists have no compunction about acting collectively and drawing boundaries against perceived attacks and enemies.) Ignoring and passively isolating provocative individuals and ideas has been the de facto course for controversy management.

Finally, in this heterogeneous intellectual space—starting with behavior genetics but extending to genetics in general—criticism has become a specialized activity effectively monopolized by a small group. As I showed in chapter 3, when the IQ controversy raged in the early 1970s, driven by Shockley, Eysenck, Herrnstein, and, above all, Jensen, the response from scientists was voluminous and diverse. I focused on figures like Lewontin, Hirsch, Dobzhansky, and Kamin, but they were but a sliver of the total. Also involved were major figures from genetics and biology as well as psychology and the social sciences, and several scientific societies issued collective statements.[50] Further, the critical outpouring was not just a singular episode driven by 1960s radicalism. Geneticists, in particular, displayed a willingness to engage in responsible deliberation before and after the IQ controversy; earlier, they had participated in collective discussions on race, and later they would debate the dangers and governance of recombinant DNA technology.[51]

By the time of the *Bell Curve* controversy, scientists' participation had radically shifted. Responses to the controversy were still voluminous, but most of them were by social scientists, historians, political pundits, and cultural critics.[52] Among biologists and geneticists there were very few commentators, and almost all of them were members of the already established group of critics—Hirsch, Gould, Lewontin, Beckwith, and Wahlsten. Computational geneticist and psychiatrist Bernie Devlin and colleagues edited a book subtitled *Scientists Respond to* The Bell Curve.[53] Devlin and animal behavior geneticist Doug Wahlsten were the only two biological scientists among the twenty-five contributors. Other collections of responses had none outside the list of usual suspects. Once considered an urgent collective matter, geneticists now considered *The Bell Curve* too "stupid" to rebut, as David Botstein quipped.[54]

Jonathan Beckwith described how the HGP's original ELSI working group wanted to organize the genome community to make a statement about *The Bell Curve*.[55] The working group believed that *The Bell Curve* misused genetic concepts and evidence but also that public pronouncements of many genetics leaders—the tendency to promote research with "astrological genetics" metaphors—would lead the public to believe incorrectly that behavioral traits are genetically determined. The group drafted a statement and submitted it to the office of Francis Collins, the HGP director. In 1996, after two years' delay, his staff allowed it to be published as a letter in the *American Journal of Human Genetics*.[56] Among the working group signatories were a handful of geneticists along with the philosophers, social scientists, and lawyers, but the group had hoped for an endorsement from the HGP leadership. According to Beckwith quite the opposite occurred: this event dovetailed with several other tensions, which resulted in the de-funding of a proposed ELSI project on the implications of behavior genetics, the resignation of the chairwoman, and Collins's reconstitution of the ELSI group's membership and priorities. Not only was the genetics establishment wary of commenting on behavior genetics, their actions encouraged ELSI efforts to stay away from the claims and actions of scientists and to focus only on downstream implications and applications of the science. *The Bell Curve* may have been too stupid to refute, but perhaps geneticists were wary of establishing a precedent.

Thus by the mid-1990s scientific responsibility had been reorganized. The practice of public commentary and critique on major genetics issues, long a widely distributed responsibility among geneticists, had become a specialized task partly delegated, on the one hand, to ELSI specialists, and partly ceded, on the other, to a small cadre of critics. In the process critical responsibility had taken on a symbolic discredit. Beckwith relates some of scientists' disparaging attitudes toward ELSI: "welfare for ethicists" making "vacuous pronunciamientos."[57] His memoir also relates a career on the receiving end of crossed arms and sideways glances for his scientific activism. When asked by a journalist about Richard Lewontin and other critics' positions on behavior genetics, Botstein said,

> Richard Lewontin is being extremely up front by saying [his opposition] is political. I have been saying for years, in private of course, that the problem with Dick and people like him is that they are politically motivated. . . . You have gotten a wonderful quote from Lewontin, which I love. . . . Lewontin has, for the first time, admitted that all this stuff is about politics and not about science. Basically he's taken himself out of the argument.[58]

Here Botstein expresses a widespread attitude that criticism of behavior genetics is a political act, something that undermines one's scientific bona fides. But if it is too "stupid" to fight on scientific grounds and attacking on political grounds is forbidden, then no wonder most scientists have left the criticizing to the veterans.

In *Exit, Voice, and Loyalty*, economist Albert O. Hirschman examined responses to decline in organizations.[59] Hirschman argued that when problems accumulate, members usually respond by heading for the doors. It is only when the exits are blocked or conditions inculcate some kind of loyalty that they speak up and try to effect change. In behavior genetics, the conditions for voice are not present. Since most researchers doing behavior genetics have other disciplinary memberships, they have ample opportunities for exit, possibly to return when the trouble blows over. Since the field lacks a disciplinary core with any gravity and requires so few investments for participation, few have developed deep loyalties. Further, the cultural scaffolding for voice is limited: In this archipelagic field, basic questions—who should speak? with what authority? using which standards? to which audiences? with what kind of recognition?—don't have clear or encouraging answers. As a result the voice of scientific responsibility has been largely quieted. Ignoring and passively isolating provocations and problems, allowing them to float indeterminately, has become the de facto strategy.

Symbolic Capital of Notoriety

In a profile of behavior genetics in *Science*, a journalist wrote, "If you're looking for a quiet life away from controversy, behavior genetics isn't going to be the field to join."[60] For some this is a burden to bear, and for others it's a reason to stay away. In "Science as a Vocation," Max Weber captured a basic image of the scientist who eschews the spotlight: toiling long hours in the lab, filled with the "strange intoxication, ridiculed by every outsider" to succeed in an experiment or calculation whose value is unrecognized beyond a small community of specialists.[61] For some researchers, the appeal of behavior genetics is precisely in the capacity it offers to break out of this image. It yields the opportunity to pursue what might be called the symbolic capital of notoriety: the ability to seek recognition among broad audiences, the capacities to speak to broad concerns as the "impresario of the subject" and to form oneself into a scientific "personality."[62]

A crucial motivation for participation in behavior genetics was to do research that everyone cares about, to be at the center of basic questions about human nature. Dean Hamer described his decision to table a successful career in basic research on cancer genetics to study homosexuality:

> People often ask why I switched from a field as obscure as metallothionein research to one as controversial as homosexuality. The answer is the same

that most scientists give for what they do: a combination of curiosity, altruism, and ambition (especially curiosity, both personal and scientific), combined with one more factor—boredom. After twenty years of doing science, I had learned quite a bit about how genes work in individual cells, but I knew little about what makes people tick. . . . I realized that even if I stuck with this research for another ten years, the best I could hope for was to build a detailed three-dimensional replica of our little regulatory model. It didn't seem like much of a lifetime goal.[63]

No longer content with the recognition of the small expert scientific community, behavior genetics enabled Hamer to reinvent himself as a scientific celebrity. In his popular book on the research, Hamer recounted the tremendous scientific and public attention he received following the "gay gene" announcement. Although he portrayed the media appearances and endless questions as a tedious chore, he was clearly excited by the attention, having written two additional books, researched what he calls the "God gene," and produced a documentary about sexual tolerance.[64]

Molecular geneticist Bruce Lahn began his career researching the evolution of human sex chromosomes. He then switched to the evolution of genes implicated in neural growth and the answers they might hold for racial differences in intelligence. "Bruce is in a hurry to be famous," said Martin Kreitman, a colleague at the University of Chicago.[65] Lahn's work has garnered copious media attention—including praise from racial conservatives—but many colleagues thought the genetic links to intelligence differences and race to be unfounded and irresponsible.[66]

A very recent example demonstrates how participants can turn notoriety into an ironic kind of virtuous circle. In February 2013, the firm BGI-Shenzhen (formerly Beijing Genomics Institute) announced a search for intelligence genes by collecting a sample of 1,600–2,200 individuals that score in the top 1 percent for verbal and quantitative reasoning skills and comparing their genomes to the general population.[67] The project was pitched to BGI by Stephen Hsu, a theoretical physicist and vice president for research at Michigan State University, and it involves a partnership with leading behavior geneticist Robert Plomin. The project, promising to deliver the "genius genes" in three months (the results are not in as of this writing in late May 2013), received splashy coverage in the *Wall Street Journal*, *Nature*, the *International Business Times*, the *MIT Technology Review*, and *Vice Magazine* among others. The partnership follows a classic pattern: Hsu the nonexpert enters the field with a bold idea, he partners with Plomin who will happily "give away" his expertise and, more importantly, his study population, while charges of eugenics and American anxieties about the rise of China swirl. Instead of starting his leadership role as a sober scientist, Hsu becomes an audacious visionary, BGI receives breathless coverage and is labeled one of the "50 Disruptive Companies of 2013," and Plomin gets his data. Plomin, the veteran at this kind of thing, sounded almost blasé, "Maybe it will work, maybe it won't . . . But BGI is doing it basically for free" and "intelligence does push a lot of buttons. It's like waving a red flag to a bull."[68]

Although Hamer, Lahn, and Hsu may be unusual (but not unique), the connection between ambition and splashy, publicly provocative topics isn't rare among behavior geneticists. One animal researcher complained that "many people who do behavior genetics harp so much on the human dimension of what they're doing even if their research probably will have very little to do with things that happen in humans." Among those studying behaviors imputed to be "aggression" in mice, "the first thing out of your mouth will be telling me what great thing you're doing for humanity because you're going to solve the problem of people beating up their spouses and why we go to war and all of these kinds of things."[69] The ASHG group of behavior geneticists described the ultimate mandate for their research as its "wide public interest and social importance. . . . Public knowledge, program design, and policy development should rest not on popular myths but on findings from the best available science."[70] The interest in connecting behavior genetics to public concerns is common.

As many other scholars have shown, behavior genetics has long been a fascination of the print media.[71] For those who desire the attention and can offer a good quote, participation in behavior genetics can almost guarantee repeated media coverage. Prominent behavior geneticists Robert Plomin and Thomas Bouchard have probably been quoted several hundred times each in mainstream and science journalism outlets. Of course, much attention is also offered critics of behavior genetics, who are often quoted as counterpoint on controversial issues.

Media sociologist Massimiano Bucchi has argued that scientists sometimes engage the public media because it affords them opportunities to make arguments that are ordinarily prohibited in formal scientific venues.[72] For example, Bouchard and colleagues on the Minnesota Study of Twins Reared Apart often discussed the improbable coincidences they

observed: separated twins who both wear the same configuration of jew-
elry, or who are both "great gigglers" sharing a disinterest in politics, or
who share the same measures for circulatory health despite discordance
for smoking. One oft-repeated example was a pair of twins, one raised as
a Nazi youth and the other as a Jewish émigré, who both

> were wearing wire-rimmed glasses and mustaches, both sported two-pocket
> shirts with epaulets . . . they like spicy foods and sweet liqueurs, are absent
> minded, have a habit of falling asleep in front of the television, think it's
> funny to sneeze in a crowd of strangers, flush the toilet before using it, store
> rubber bands on their wrists, read magazines from back to front, dip buttered
> toast in their coffee . . . [etc.][73]

Although the researchers described such coincidences as evidence of
genetic effects in the media, scientific audiences would not accept them
as proper data.[74] In popular media scientists can also speculate openly
on their research. For example, Richard Ebstein, a molecular geneticist
studying the connection between "risk taking" and the dopamine $D4$ receptor gene, told a reporter, after John F. Kennedy Jr.'s 1999 plane crash
off of Nantucket, that the Kennedys "undoubtedly bear" the "novelty-
seeking and risk-taking gene" even though he "obviously [hadn't] tested
them."[75] His behavior genetics research got him a call from a reporter
looking for a quote, and while he could have demurred, he indulged the
impulse to make a brash, unsupported claim.[76] With journalists eager for
good copy, a behavior geneticist need not retool a career like Hamer or
swim in the dangerous waters of racial research to get a little taste of
notoriety.[77]

Behavior genetics' symbolic capital is only problematically linked to
economic or political power. Certainly some have been able to convert
their capital as behavior geneticists for other forms of recognition. One
example is Glayde Whitney who used his BGA presidency to assert the
genetic inferiority of black people and then became a writer for extreme
right-wing political causes. One animal behavior geneticist said, "Whitney made much more impact by starting saying these things, and got
much more attention than he ever got [for his mouse research]. And he
must have gotten a lot of kind of satisfaction, a lot of people telling him,
'You're so courageous and doing the good thing.' Okay it's [white supremacist politician] David Duke who says it, but somebody says it."[78]
Likewise, Rushton became the president of the race-research funding

Pioneer Fund.[79] Herrnstein and Jensen have certainly received recognition of similar types (and the sales of trade books) without making such extreme conversions. Hamer's analogous recognition has come from the left, which has sometimes advanced the argument that genetics proves homosexuality is not a choice and should therefore be a protected status.[80]

Such conversions of behavior genetics' symbolic capital are not common. Behavior geneticists have generally not found their expertise sought, directly at any rate, by policy makers. Psychological behavior geneticist David Rowe expressed a common complaint that behavior geneticists lack a place at the policy table.[81] Critics have charged behavior geneticists with using notoriety to attract research funding, but the effects may have been the reverse.[82] Principals of the widely publicized Minnesota Study of Twins Reared Apart always complained that publicity and biased reviewers hindered their ability to secure sufficient funding from public sources and that money from the Pioneer Fund (and other private sources including conservative billionaire David Koch's foundation) was necessary to continue.[83] Money from private foundations, especially politically oriented ones, has always been uncommon in the field outside the racial provocateurs. So too has money from pharmaceutical or biotech corporations, though sometimes this shows up in more psychiatry-oriented research.[84] Several behavior geneticists I spoke with felt that the field's notorious reputation in general and the media coverage they themselves had received had led to grant applications being denied—perhaps a norm of scientific humility was being enforced.

Notoriety has had ambivalent effects within science too. On the one hand behavior genetics is sometimes a path into in high status journals. For example, Dean Hamer and Bruce Lahn have published in excellent journals, but their work on behavior and for Lahn, race, has appeared in *Science*. Psychologists' research rarely appears in *Science*, but many psychological behavior geneticists have appeared there. Notoriety is dangerous too. One psychological behavior geneticist explained, in reference to Rushton, how one can get trapped by provocative claims, "Whether or not you're right or wrong, you'll have to defend it, because that's how you'll be seen."[85] Intelligence researcher Robert Sternberg has written, "I am at a loss as to why Jensen persists in studying the problem of black-white differences. . . . I . . . wish Jensen would make better use of his considerable talents."[86] And more dramatic was the response to James Watson's comment in an unguarded moment during an interview with London's *Sunday Times Magazine* that he was "inherently gloomy about the prospect of Africa . . . all our social policies are based on the fact that their intelligence is the same as ours—whereas testing says not really."[87] Watson apologized "unreservedly" and stated that there was no scientific basis for his remark, yet he was forced to resign his directorship of the Cold Spring Harbor Laboratory, which effectively ended his eminent career.

Behavior geneticists often complain that when they are accused of making provocative public claims, reporters who don't understand the science and are themselves eager to sensationalize have misquoted them or misinterpreted their research.[88] There is a need, clearly, to distance oneself from making provocative claims—blaming the media provides a level of deniability that acts as a safety valve. These kinds of pressure release make it possible to play with the symbolic capital of notoriety without becoming consumed by it. Notoriety is a paradoxical symbolic capital: one must be bold and full of conviction to earn it, but also cautious to avoid its dangers. For the most provocative individuals, the tradeoffs are fairly clear—they exchange notoriety and broad, especially public, attention for respectability. But the complexities, ambiguities, risks, and limits in exchanging the capital of notoriety for economic, political, or scientific capital suggests that something other than instrumental success in these realms is at stake.

In *The Quest for Excitement* social theorists Norbert Elias and Eric Dunning analyzed the modern appeal of sports as a reaction, on the one hand, to the diminution of expressive violence and the cultivation of self-restraint, and on the other, to the increased routinization of daily life.[89] Sports offers opportunities for the cathartic release of emotion and the dramatization of risk and violence otherwise lacking in the lives of modern people. There are few areas of life as rationalized as science. The passion in Weber's scientific vocation is built on self-restraint, specialization, and obscurity. Behavior genetics allows scientists some release from the strictures of the vocation without leaving the domain of science. Thus the symbolic capital of notoriety can be an end in itself, a way of using science to pursue the quest for excitement.

Tendrils of Geneticization

The analysis thus far has laid to rest the idea that responsibility is a matter of individual commitment. Now we turn to the broader social and cultural effects of behavior genetics: the rise and impact of "geneticization"—the increasing definition of human behaviors, diseases, capacities, and differences in genetic terms.[90] Here too, focusing on individuals has obscured matters: pernicious geneticization has too often been understood either as the product of individual provocateurs overstepping the limits of science or individual nonscientists who ignorantly overinterpret implications. Some critics display the opposite tendency, elevating geneticization to an all-powerful cultural force. Barnes and Dupré's "astrological genetics" gets at the ways that genetic determinism prevails as an explanatory frame in public discussion even when context or environment are mentioned.[91] Nelkin and Lindee go further, tracking so many examples of geneticized discourse in popular culture one might forget that human life is ever described in other terms.[92] Neither the individualistic nor the broadly culturalist view captures the specific logics of geneticization in behavior genetics. The aim here is to show that geneticization is the outcome of the structured intellectual logic of behavior genetics driven by its associated social logics of assertion and silence. This directs our attention both to the institutional weakness of behavior genetics and to the various ways its ideas can become entangled with social life, from individual performance to the political culture of neoliberalism.

The idea of geneticization is derived from medicalization: the extension of medical authority and definitions into new domains of social life. According to sociologist Paul Starr, the history of the US medical profession is the meteoric rise of its cultural authority through its growing capacities to monopolize definitions of health and illness, to institutionalize those definitions, and to control access to the institutions.[93] In an era of HMOs, alternative medicine, and antivaccination crusades it is easy to forget how utterly hegemonic the cultural authority of the medical profession still is. Psychology's cultural authority is of a different sort. According to sociologist Nikolas Rose, psychology has eschewed medicine's strategy of control and succeeded through "generosity," putting its concepts and tools in the service of important institutions (military, education, industry, for example) and distributing them widely in popular culture (self-help books, for example) with little concern for how they are used and altered.[94] As a result psychological concepts are inescapably part of how we Western people conceive of our own minds and selves.

The cultural authority of behavior genetics pales in comparison. Certainly the geneticization it fosters has become widely available via the media, but its institutionalization is otherwise extremely thin. Beyond the few clinical psychologists and psychiatrists who might deploy its ideas, behavior genetics thus far has no clear foothold in medicine, education, criminal justice, industry, or policy making. Although its implications for these domains are hotly debated, the field has no capacity to impose definitions through these institutional channels. Many tenets of medicine and psychology have become part of the rarely questioned cultural infrastructure of daily life. The same can't be said of behavior genetics; indeed, though its basic ideas are widespread they are also regularly criticized and challenged when publicly presented. Further, many alternate discourses and practices circulate in the cultural sphere that challenge the determinist suppositions of geneticization. For example, the whole American culture of self-help is predicated on the idea that we can change our behavioral dispositions. Policy scholar David Kirp has described a growing movement of scholars, activists, and policy makers seeking to spread high quality preschool programs to the poor to change kids' intellectual and social trajectories.[95] Such a movement rejects the idea that high heritabilities for IQ and behavioral problems suggest schools are "good enough" to unlock children's genetic potentials. Geneticized ideas float in a diverse cultural soup of competing ways of thinking about behavioral potential and difference.

Even with its limited and contested cultural authority, behavior genetics ideas can have serious effects. Social psychologists have identified the phenomenon of "stereotype vulnerability" as the anxiety experienced when a person believes their performance might confirm a negative stereotype about a group to which they belong. The classic examples concern testing situations. When told that a test measures inherent ability, African Americans (reacting to the stereotype that they are less intelligent than other groups) and women (reacting to the stereotype that they are poor at math) perform worse than whites and men—*and* worse than those who take the same test but are told something less threatening about the test.[96] In the domain of parenting, psychologists have shown that parents who believe that parenting doesn't much matter to kids' outcomes tend to have kids with worse outcomes.[97] It has also been argued that a significant barrier to the identification of effective interventions to boost achievement is social scientists' pessimism about the prospect of interventions.[98] Thus some of the ways that behavior genetics represents the world can become self-fulfilling prophecies.[99]

Another effect has been to change understandings of social responsibility in politics and law. For example, Dorothy Nelkin tracked the use of behavior genetics concepts in a variety of political debates about social policy.[100] These arguments were invoked to undermine the idea of collective responsibility and also confidence in the prospects that social programs might solve problems. Historian Jonathan Harwood was slightly more skeptical about the causal impact of behavior genetics ideas in social policy discourse.[101] He argued that claims of Jensen, Herrnstein, and others in the 1970s did not lead to the demise of President Johnson's Great Society vision, but rather gave policy makers, who for a variety of reasons were already abandoning it, a set of post hoc rationalizations and legitimating arguments. More recently, gay rights activists have used genetic determinist ideas to justify their claim that homosexuality is an inherent disposition (not a lifestyle choice) and thus worthy of equal rights.[102]

Although the power of legitimating discourse should not be minimized, scholars recently have sought to test more directly the effects of behavior genetics ideas. Lisa Aspinwall, Teneille Brown, and James Tabery recently conducted an experimental survey of sentencing judges to understand the effects of biodeterminist ideas. Judges were given a vignette about a hypothetical crime where the perpetrator had been diagnosed as a psychopath.[103] Half the judges were presented mock expert testimony explaining the psychopathy had a genetic-biological cause, and for the other half, expert testimony didn't mention biology. Judges, it turned out, viewed genetics as a mitigating factor and on average reduced the sentence from about fourteen to about thirteen years. In calling for reduced sentences, judges seemed to be thinking more about genetic determinism in terms of culpability rather than future harm to society.[104] In a different study, sociologist Sara Shostak and colleagues used a national opinion survey to look at people's views of genetic causation.[105] They found, first, no evidence for the common idea that whites, conservatives, and the socioeconomically privileged are more likely to view genetics as determinative of health or social success. But they also found that people who did view genes as strongly determinative also supported policies involving genetics (such as support for the Human Genome Project, more genetic tests, and disclosure of hereditary conditions). These studies focus on a narrow set of policy topics, yet they suggest that a broad narrative of geneticization and astrological genetics can indeed shape the opinions and actions of professionals and ordinary people.[106]

Much ethical reflection on behavior genetics concerns science fiction scenarios; for example, if parents could genetically enhance embryos to be hyperintelligent, should it be allowed?[107] Behavior geneticists often imagine their science might one day lead to pharmaceutical behavioral intervention. In reality, the science is extremely far from delivering such therapies, but other relatively mundane applications could, in principle, be applied today. Take Caspi and Moffitt's studies about interactions between genes and environment, which show that children with particular genetic variants are especially likely to become antisocial or depressed when they experience abuse or traumatic experiences growing up.[108] Many social support services for poor people are severely underresourced. Child protection services in many cities face chronic shortages of sufficient quality foster housing and adoptive parents, and caseworkers often face an impossible caseload. Could genetic screening help triage the system? Perhaps studies like Caspi and Moffitt's might identify genetic factors (among others) to identify the children at greatest risk to suffer negative outcomes. Those with the risky alleles might be given more intensive casework or bumped up waiting lists relative to those with the more "resilient" genotypes. Clearly a fully funded child welfare system is preferable, but perhaps behavior genetics information could help an insufficient system better reflect a utilitarian ideal.[109]

In principle, there are few barriers to implementing this system. Testing individuals for specific genetic variants has become cheap and easy in this era of personalized genomics. And foster care agencies already use multifactorial systems to monitor children's risks and aid decision making. The main issues would be deciding how well established a genetic association has to be to act on it—replications of these claims have been ambiguous—and how to weigh genetic risks among others. These decisions would, of course, be made in a public culture where the legitimacy and authority of social services is crumbling and genetic science has been heralded as key to solving medical and social problems. I present this possibility not to advocate for behavior genetics to be a part of social policy (indeed I would be against it), but to suggest that from a technical standpoint this *could* happen today.

The "rhetoric of reaction," according to Albert O. Hirschman, tries to links social problems and solutions to "perversity," "futility," and "jeopardy."[110] Hard-line eugenics from the early twentieth century certainly fit the mold. Genes separate people into distinct types; genetic determinism means social interventions will either be futile or have perverse consequences (encouraging the "unfit" to propagate); and failing to address the immigration and fecundity of the unfit relative to the "better classes" jeopardizes social order. Contemporary behavior genetics has long been accused of fitting a similar logic, but this seems incorrect. Rather than a discourse of paleoconservatism, behavior genetics strikes chords of neoliberal discourse.

Neoliberalism, according to theorist David Harvey, is a political and economic theory that claims human flourishing can best be cultivated by "liberating individual entrepreneurial freedoms and skills within an institutional framework characterized by strong private property rights, free markets, and free trade."[111] Beyond securing these conditions for economic life, any efforts by the state to promote social welfare will inevitably lead to the suppression of liberty and individual potential. The state can do better by extending markets into realms like education, health care, and environment. Politically, neoliberalism is often driven by a quasi-populist critique of state and professional experts as self-serving elites who must be disciplined by the market (which is to say, replaced with economic or corporate experts). Behavior genetics' affinity with neoliberalism can be seen in two basic areas.

Behavior genetics, the science of individual differences, valorizes individuality along similar lines as neoliberalism. Margaret Thatcher once quipped, "There is no society, just individuals and their families." Most behavior genetics studies look for the roots of individuality in genetic and family effects; usually the effects of "society" are an out-of-focus remainder. Psychological behavior geneticist David Rowe wrote,

> When men and women are allowed to explore social roles over a wide social range, they may pick ones better suited to their *individual* biological dispositions; that is, only when there is freedom of choice can men and women make choices truly diagnostic of their biological proclivities. If people can find greater happiness in "niche picking" . . . then permitting "biological determinism" to flourish may be evaluated more positively than social engineering programs of either the radical left or the radical right.[112]

Rowe, a political conservative, echoes neoliberalism by viewing social life as something like a market where people are consumers choosing among lifestyles. More choices and more freedom will solve problems that states have tried to solve through "engineering" social outcomes. Now consider the views of an animal behavior geneticist and deep critic of Rowe's brand of behavior genetics:

> Each of us is an absolute unique experiment in this world, and with that kind of uniqueness there are certain social implications that come from that. . . . And I must fall more to the left [politically] because if biological individuality is a fact, then in fact we need to create societies that maximize individual liberty and freedom and opportunity.[113]

Despite being bitterly opposed scientifically and voicing opposite political commitments, this speaker echoes the same central message. Certainly Rowe, the conservative, emphasizes choice and the fear of central planning, while the liberal speaker mentions opportunity, but both these behavior geneticists similarly valorize individuality and its cultivation through the promotion of social freedom and liberty. Contrast, for example, the language of disparity and inequality that animates public health and sociology.

Behavior genetics' second area of affinity with neoliberal discourse is in its populist-inflected critique of expertise. Behavior geneticists have long contrasted the commonsense knowledge of regular people over the confusions and obfuscations of the psychological establishment. Sandra Scarr said, "parents of two or more children know perfectly well that their children are different for reasons that have nothing to do with their training regimens."[114] Rowe made the point more explicitly political by claiming, "if social scientists come to accept these conclusions [of the limits of parental influence], the idea that *the way academics raise children would really be best for everyone* must be abandoned as well."[115] This claim echoes an implication of Jensen's critique of "compensatory education"; it didn't help kids, but it kept a lot psychologists and educational "experts" in work. And behavior geneticists often told me that their research helped people cope with mental illness because it swept away clinicians' tendency to blame problems on "refrigerator mothers" and "bad parenting."[116] Behavior geneticists are attacking more than scientific ideas, they are charging the (environmentalist) scientific establishment with being less interested in helping people and solving problems than reproducing their own position and authority and propagating their own elitist values. This isn't, however, a populist rejection of expertise. Neoliberalism criticizes the expertise of teachers, social workers, and civil servants as more oriented to self-dealing than problem solving. The answer is removing the authority of these "corrupt" experts and replacing them with economists and corporate managers. Likewise, the behavior genetics critique seeks to replace social authority and its hidden, self-serving politics with objective, apolitical biological authority.

Behavior genetics is not an economic discipline, but the arguments for its authority parallel those of neoliberalism. And its broad strokes image of the flourishing individual fits the neoliberal prescription for social order. The relationship is mutually reinforcing. Behavior genetics' orientation is informed by neoliberal politics, but it also generates a view of human nature that supports those politics.

Conclusion

We tend to think about ethics and responsibility in individualistic and voluntaristic terms. Being ethical or responsible is about an individual commitment to doing the right thing even when it is inconvenient or dangerous to do so. Seeing behavior genetics linked with a stream of provocative and mundane "astrological" claims—a pattern that many behavior geneticists find discomfiting—has produced suspicions and accusations of irresponsibility. The implication is, frankly, that behavior geneticists are bad people.

I have shown that this is the wrong way to look at the situation in behavior genetics. Definitions of responsibility and practical capacities to act responsibly are closely linked to a range of social arrangements. I showed, firstly, that there has long been a conflict among behavior geneticists about how to define responsibility in relation to intellectual freedom. The "bunkerization" of the field in response to episodes of controversy narrowed the way behavior geneticists defined responsibility. Put differently, behavior geneticists came to see "responsibility" as a weapon their critics wielded, and they came to see the staunch defense of intellectual freedom as a courageous form of responsibility to each other, the ethos of science, and the liberal values of a free society. I showed, secondly, that when "responsibility" became defined as participation in public communication projects, behavior geneticists happily complied. But these events were often judged failures; on the one hand, they couldn't overcome conflicts and often exacerbated them, and on the other, they tended to produce middle-ground statements—genes *and* environment matter—that may have helped inform and calm the public, but which satisfied few experts.

I also argued that responsibility is linked to field-level forces, in particular capacities for authority and control and the reward structure. As an archipelagic field, behavior genetics has few barriers to entry, multiple groups of actors with little authority over each other, and little sense of mutual accountability. It is easy to make bold public pronouncements on behalf of behavior genetics but very difficult for other behavior geneticists to regulate this activity. Such regulation has come to be viewed as "criticism," an increasingly stigmatized and specialized activity among self-identified behavior geneticists and among biological scientists more broadly. The field's structure makes it easy to duck or flee problems, leaving them to the seasoned critics. Further, behavior genetics encourages irresponsible public communication by enabling the pursuit of what I have labeled the symbolic capital of notoriety. Notoriety is earned by deploying behavior genetics in public to speak a bit provocatively about human capacities and differences. It's dangerous. Too much can totally transform or undermine one's scientific position. But a skillful use can open up new public *and* scientific opportunities while offering a burst of excitement in the ordinarily staid scientific life.

Critics and ethicists have worried about responsibility in behavior genetics because they fear the rise of astrological talk and geneticization. Sociologist Adam Hedgecoe has criticized geneticization as a polemical, political concept; empirical evidence that genetic ideas are more prevalent in media or in public perceptions is at best mixed.[117] Clearly, behavior genetics lacks the kind of institutional foothold that would give geneticization powerful cultural authority. Again the field is implicated: the same weakness and dispersed organization that proliferate geneticized discourse also undermine the potential for institutionalization that might strengthen it. It is worth remembering that not all geneticization arises from behavior genetics, and also that its plausibility has been bolstered by the growth and success of genetics generally. I gave several examples that show the tendrils of geneticization need not change institutions, media, or opinion writ large to change practices, lives, and policies. Finally, the political fear that geneticization reinforces reactionary conservatism misses behavior genetics' affinities with neoliberalism. Neoliberal concepts of individuality and potential find support in behavior genetics ideas and vice versa. Behavior geneticists' quasi-populism finds inspiration in neoliberalism's politics of expertise. And, indeed, the field of behavior genetics might give us an idea of what neoliberal destructuring of science might look like.

So behavior geneticists are not irresponsible people, but neither is the problem due to the inherent complexity of ideas or the simplemindedness of nonscientists. The possibilities for responsibility are intimately connected to the history and structure of the field. This makes change much more than a matter of individual commitment. Behavior genetics is in many ways an idiosyncratic field, but some of the forces affecting responsibility there affect science widely. For example, symbolic capital from publicity is becoming an increasingly important resource for scientists.[118] Avowals of scientific responsibility are becoming detached from practices of critique.[119] ELSI projects have struggled to engage scientists.[120] And calls abound for abandoning the disciplinary organization of science.[121] Thus behavior genetics has lessons for thinking about misbehaving science more broadly. To these we now turn.

Misbehaving Science

Behavior Genetics and Beyond

Before drawing out the implications of this analysis, it will be help-ful to recap the story of behavior genetics' development. The sci-entific and political embarrassment of eugenics and scientific racism in the first half of the twentieth century had stigmatized scientific interest in the genetics of behavior. In the late 1950s and 1960s, corresponding to the postwar expansion of science, a group of scientists worked to reha-bilitate the topic. Led by animal behaviorists, they began to assemble a disciplinarily and intellectually diverse research network. To ensure the field would be firmly scientific and avoid becoming ensnared in politics, they agreed to shut racial comparisons and eugenics out of the agenda and excluded scientists known to have provocative aims. Further, animal and human researchers formed an implicit pact. They would build and defend the field together, and animal research would serve as a model for human research, helping it to stay objective, biological, and avoid politi-cal entanglement. The end of the 1960s seemed the dawn of a golden era; a diverse and enthusiastic group of researchers was poised to explore a wide-open intellectual domain.

In 1969 a crisis struck. Arthur Jensen drew widely on behavior ge-netics in an article claiming that genetic differences explain the IQ gap between blacks and whites, and educational efforts to close cognitive gaps were doomed to fail. The article ignited a political and scientific firestorm. Hundreds of scientific and popular articles were written in re-sponse. Activists and students protested and disrupted behavior genetics events. The field found itself dangerously at the center of debates about civil rights; the effort to insulate it from politics had failed. What would behavior geneticists do? A small group wanted to drive out Jensen and the brand of research he represented and recover the scientific vision of the 1960s. But the field's critics didn't see Jensen as a bad apple; they claimed that behavior genetics was inherently bad science—determinist, fatalist, racist, classist. Behavior geneticists circled the wagons around Jensen. It seemed that defending him was the best way to save the field.

This decision set into motion a fateful reorganization of the field during the 1970s, away from an emerging transdiscipline and toward its fragmented archipelagic form. First, its disciplinary character shifted as scholars from the social sciences, but especially genetics and the biological sciences, began to dissociate from the field. Entering the field were clinical research psychologists who study humans. Second, behavior geneticists became concentrated in a small number of universities, their network tightened, and they adopted a "bunker mentality," which discouraged criticism within the group, to defend collectively against the attacks that still buffeted the field. Third, animal researchers lost power in the field as their numbers were diluted, their disciplinary allies grew distant, issues surrounding the IQ controversy dominated the agenda, and voicing their views could be interpreted as disloyalty. As their influence waned, so too did their vision of the field as a robust, intellectually heterogeneous, transdiscipline where scientists from different backgrounds would vigorously engage each other to uncover basic principles of the genetic inheritance of behavior. What prevailed instead was the much more casual, minimalist, and instrumental vision of the field held by the ascendant human researchers. They saw the field less as a new, open-ended discipline and more as a safe haven in which to develop tools in order to launch attacks on their home disciplines. Rather than an end in itself, behavior genetics became seen mainly as a means of competing for recognition in other disciplines, particularly psychology and psychiatry.

By the 1980s behavior genetics was bruised and beleaguered; the long IQ controversy had stigmatized the field and undermined its authority. How could so troubled a field be a means for making scientific reputations in other disciplines? Behavior geneticists learned how to turn controversy to their advantage. One strategy was to vigorously attack established positions and ideas in psychology—advancing a genetic explanation for phenomena that had been understood in other ways. Each round of battle elevated the importance of behavior genetics and spurred the productive activity of defending the method, collecting data, and making new claims. Another strategy was to form partnerships with open-minded scholars and to use the alliance to generate new productive conflicts. This process never settled fundamental disputes about behavior genetics, but it allowed participants to build meaningful careers.

One critique of behavior genetics was that its core methods made genetic inferences without actually measuring DNA. By the 1990s, technological advances spurred by the Human Genome Project offered tools to correlate behavior directly to DNA. A new era was heralded, one that would finally put behavior genetics on firm scientific foundations. As molecular geneticists newly flooded into the field, they didn't act like sober, objective experts but made wildly speculative, determinist, and "astrological" claims about reading individuals' fates from their genes. Scientifically, they were rapidly and repeatedly embarrassed as it turned out that genes for behavior were nearly impossible to identify definitively. The hope that these tools would transform the field was disappointed. Molecular tools became widespread among the islands, but they could not reverse the field's fragmentation. Indeed, they spurred further competition and mistrust, encouraging veteran behavior geneticists to position themselves against the newcomers. Interestingly, they represented themselves as antireductionist defenders of holistic understandings of people and behavior—terms they criticized in social scientists a short time earlier. Behavior geneticists reinvested in the old-fashioned, nonmolecular methods. And rather than molecular tools fundamentally changing the way behavior geneticists work, those tools were bent to prevailing styles of research. They were used to accumulate disputed results rather than as a tool for articulating experimentally, developmentally, and physiologically how a gene affects an organism's behavior.

The final chapter built on the preceding analysis to explain why behavior geneticists' style of public communication has so often been provocative, deterministic, "astrological," and political. Through their long history of controversy, behavior geneticists came to see their scientific responsibility as the zealous defense of scientific freedom, not the management of misleading communication to the public. They have dutifully participated in projects run by ethicists to rationalize communication about the science, but these have ended up reproducing the disagreements they had aimed to overcome. The field's chaotic organization makes its ideas easy to appropriate provocatively and makes it difficult for scientists to affect each other's behavior. Furthermore, the basic reasons many scientists participate in the field include the capacities it af-

fords to gain public attention and notoriety and turn these into professional assets. "Irresponsible" communication is almost guaranteed.

Throughout the argument, I showed how these changes in the field affected, and were often reinforced by, scientific commitments, agendas,
practices, and interpretations. Here are some of the highlights. Behavior
geneticists, uneasily and at arm's length, endorsed a partial genetic explanation for racial differences. Heritability became (through countless
variations) the field's central scientific principle. Although considered by
geneticists (and many of the field's founders) a simple descriptive statistic of limited value, it became for behavior geneticists a protean resource
to manufacture ideas about social structure, individual fates, the origins
of traits, the irrelevance of social environment, and then the action of social environment. Genomic tools were also assimilated into this framework, which assumed that the genetic inheritance of behavior could only
be understood by conceiving the forces of nature and nurture as distinct,
and finding ways to measure them separately. Alternate ways of conceiving the problem became separated from one another; for example, connections were hardly pursued to ideas of social evolution or behavioral
ecology, physiological and neural development, or between animal and
human behavior. These and other frameworks have developed in isolation, and others were never pursued. Finally, these ideas and patterns
of public communication have spread the tendrils of geneticization. Although poorly institutionalized, these ideas can have broad effects: from
warping individual psyche and performance, to the legitimation and execution of social policies, to sustaining symbiotically broad logics of political culture.

The puzzle that organized this book was "misbehaving science": How
could behavior genetics sustain persistent, ungovernable controversy for
fifty years and counting? How could polarized dissensus on its epistemic
status endure? The answer is that controversy became inscribed in the
social and cultural logic of the field. Behavior geneticists were able to define a version of the field that enabled them to craft controversy into viable scientific careers.

The questions for this conclusion are: How can we go beyond the spe-
cifics of behavior genetics? What are the broader lessons of misbehaving
science? How do they help us understand contemporary science? First,
I consider some of the analytic contributions of field theory for thinking about scientific practice and controversy. Next, I recap the forces that

made behavior genetics a misbehaving science and show how we might look for these forces in other scientific fields. I apply these specifically to two trends in the contemporary governance of science, the promotion of postdisciplinary science and the outsourcing of ethical debate. Finally, I argue that a critical sociology might focus on the characteristics of science that open up and foreclose intellectual possibilities.

Field Analysis and Controversy

One of the contributions of this work is to show how field analysis might be an important tool for sociologists of science. Within science studies Bourdieu has been largely persona non grata for his polemics against the field, for his normative and rationalist descriptions of science, and for a particular image of the scientific field as a little world operating with its own logic.[1] It is this last point that particularly galls readers, I think. Science doesn't look like that, they say. It is worldly and embedded, not autonomous; so much more drives scientists than the accumulation of "scientific capital"; their interactions aren't only competitive.

The story I've told about behavior genetics has shown that "fieldness" matters. One can't explain why behavior geneticists have defended genetic explanations of racial differences, said parents don't matter, accumulated heritability findings, dismissed criticism as political, partly embraced molecular genetics, and repeatedly provoked the public without understanding behavior genetics as an evolving, twofold structure: a field of forces and a field of struggles mutually shaping each other. Crucially, behavior genetics doesn't look like the stereotypical image of the Bourdieusian field at all; it is a fragmented, shifting, disorderly archipelago, not a well-bounded, tightly organized little system. In other words, I've endeavored to demonstrate via this example how to think of field theory as an analytic strategy, not a substantive theory of science. "Fieldness" is an empirical question and a strategic location to look at scientists' practice, not a particular image or form of science to impose.[2]

Critics accuse field theory of reductionism; it's all about a zero-sum competition for capital.[3] Sociologists of science instead consider many dimensions of scientific practice: epistemic cultures, technologies or experimental systems, judgment, institutions, and politics, for example.[4] My analysis of behavior genetics drew on all of those dimensions, among others. But it also made clear that none of them in isolation could explain behavior geneticists' actions and commitments. In fact, only by linking them to the evolving forces and struggles constituting the field could I show how they shaped behavior geneticists' actions. Field theory turns the charge of reductionism on its head; it has allowed me to avoid reducing behavior genetics to either politics or epistemology (an opposition that has structured debate for over forty years) or to the network of associations, cultural cognition, material practice, organizations, or political economy.

The field framework also directs attention to an often-neglected pragmatics of scientific practice. By this I mean that scientists may adopt intellectual positions not, or not only, for epistemic, technological, or political reasons, say, but because they make the most sense given the logic of the field. These pragmatics will, in most cases, be oriented toward securing the greatest room for maneuver or best chances of continuing on a trajectory given the current state of affairs. Thus I argued that these kinds of motivations, often implicitly, were behind the defense of Jensen in the IQ debate, human behavior geneticists' preferences for a nondisciplinary field in their conflict with animal researchers, and their bet-hedging responses to disappointments of molecular genetics. It is a truism that science must be understood in context. Field analysis gives a specific definition of context: the forces and struggles structuring a particular set of possibilities for action and speech. Attention to this context highlights the paradox of the pragmatic mobilization of rationality and ideology—things that in principle are supposed to be free of pragmatic considerations.

Bourdieu always emphasizes power and domination as the engines of social action. Scientists, he says, seek a monopoly on the definition of scientific capital. The pragmatics I'm talking about still concern social power, but less domination and monopoly. Rather, they are about securing the conditions for the ongoing production of scientific capital and its conversion into various forms of symbolic and material "profit." Thus, for example, behavior geneticists often blustered about routing the flat-earther gene denialists, but the ongoing conflict allowed them to keep the factory of twin studies and variance partitioning profitable indefinitely. An eager media then helped transform these products into public notoriety, which, in turn, helped stoke the controversy.

The field approach has also opened up new ways to think about scientific controversy. Traditionally, sociologists have understood controversies as discrete intellectual contests among scientists or disputes where science and the public meet. Controversies are temporary breaches of social order where authority is at stake, so they are key sites for understanding the social processes by which scientists restore order and draw boundaries between science and society. I proposed the idea of misbehaving science to analyze controversy as a way of life, the institutionalization of anomie, in behavior genetics. I turn to this in the next section.

The field approach highlights a second kind of controversy: disputes where the forces and struggles structuring the field are at stake, instead of a direct focus on substantive issues as in the traditional approach. The controversies over substantive matters in behavior genetics (intelligence, crime, personality, mental illnesses, alcoholism, and such) have been highly repetitive, slight variations on the same polarized positions.[5] I focused on controversies where substantive disputes were the fulcrums of field transformation. Rather than reproducing polarization, they showed, among other things, how the polarization was produced, maintained, and coped with. These controversies succeed each other in time, each setting the conditions under which the next will be played out. Thus rather than one-off events, these controversies give fields a history.

Third, controversy can be a productive force, not just a disruption that imperils scientific authority. Twin studies and molecular associations are protean scientific resources not because they settle scientific questions but because they provoke conflicts that can churn endlessly. Attacks on twin studies stimulated research to defend them. When heritability claims were met with a shrug, behavior geneticists provoked conflicts with new audiences. The "rational" response to the failures of molecular tools to identify "genes for" particular traits would be to abandon and rethink the approach, but instead scientists have repeatedly raised the bet. Such controversies certainly pose risks to scientific authority. But depending on how they are embedded in the pragmatics of the field, they can spur productive activity. Having something to do might be better than having everyone believe you're right.

Finally, controversy can be a source of forms of symbolic capital like publicity and notoriety. Being controversial, getting attention, making an impact, and picking fights are all motivations that can animate scientists. Again, whether such pursuits make one a figure of respect or ignominy depends on the state of the field. Seeking notoriety might endanger one's scientific authority, or it might be a risky way to leapfrog the long process of building a sober and respectable career. Notoriety might make us rethink scientific authority. Is it solely based on rational legitimacy as Weber claimed, or as sociologist Richard Sennett has suggested, might authority depend as much on images of strength, charisma, and bravery?[6] Sociologists now focus on how justice, ethics, and virtues animate scientific practice.[7] Notoriety, celebrity, infamy, and such items from the other side of the moral ledger should be next.

Misbehaving Science: Centripetal and Centrifugal Forces

The extended case study of behavior genetics has allowed me to advance and explore the idea of *misbehaving science*, which I have defined as persistent, ungovernable controversy due to relative disorder and partial anomie. Misbehaving science is a collective problem for scientists because it threatens their authority, undermines their legitimacy (and even their morality), and coping with it consumes tremendous energy. Misbehaving science need not be utterly destructive; after all, it is a chronic not a catastrophic problem. Adaptations to it set into motion forces that tend to reproduce the conditions of controversy. These "internal" problems generate parallel issues for society. How should controversial knowledge be evaluated and used? Which experts are to be believed? Can we trust science?

The previous chapters showed how behavior genetics became and remained a misbehaving science. Allow me to draw the main strands together here before considering how this concept helps us think about contemporary science more broadly. Much of the misbehavior is rooted in the anomie spurred by the field's archipelagic structure, the way its members are organized into a set of disciplinarily identified, ambivalently associated islands. This means that behavior geneticists are guided by competing sets of disciplinary standards. Cognitively and scientifically this means they have different kinds of training, favored methods, questions, and assumptions. There is a lot of overlap enabling frequent collaborations and exchange, but, at the same time, the differences sow distrust and inhibit mutual judgment. Institutionally, it means that behavior geneticists are dependent on other disciplines for their employment and students. Archipelagic fragmentation also leads to inconsistent communication and affiliation. There are several societies and journals that focus on behavior genetics, but each is associated with a different island. There aren't any collective institutions that work to gather behavior geneticists from competing perspectives under one tent. These dimensions of fragmentation produce contradictory and ambiguous rules of practice.

A crucial upshot is that the recognition-seeking game is organized centripetally. Most behavior geneticists are less interested in battling for each other's recognition than battling for esteem in their home disciplines and other topically organized research communities. They often believe they can take other behavior geneticists' recognition more or less for granted and would rather cash in scientific capital earned there for other forms of symbolic capital. Interestingly, behavior genetics has been, to date, weakly institutionalized in most extrascientific realms. So, in sharp contrast to fields like molecular genetics or economics, it has not been a particularly reliable means for pursing money or political or social influence. The exception, of course, is fervent cultural and media interest in behavior genetics, thus behavior genetics has enabled the vigorous pursuit of the symbolic capital of media recognition, public attention, and notoriety.

These forces combine with others to undermine field participants' attachment to the field and senses of mutual responsibility. For example, it has often been possible to make scientific contributions without deep investments of energy or resources, and many of the methods exist in a portable, widely applicable form. There have been few barriers to entering or leaving the field; one need not identify as a "behavior geneticist" to participate and one can depart without abandoning much "sunk capital." This easy mobility, combined with the subtle stigma attached to behavior genetics in many quarters, hinders the development of loyalty. Thus individuals are less likely to worry about how their actions affect others or to feel responsible for how others act.

Behavior genetics has developed a regime of policing that acts to maintain individual scientists' freedom of maneuver. Most of the active policing is directed "outward" at the field's external critics. Behavior geneticists' senses of mutual responsibility are greatest when someone is under attack. Most "inward" policing is directed at would-be police. That is, behavior geneticists who would criticize others in the field face serious discouragement from their peers. Policing is in effect mobilized to promote noninterference and individual freedom, not to promote intellectual critique. Behavior genetics research certainly goes through peer review, but there is pressure to restrain skepticism and stick to technical matters. These tendencies are linked to the field's archipelagic structure, but they are more directly the product of efforts to cope with controversies. General anomie is, ironically, enforced through this set of rules.

What were the consequences for behavior genetics of being a misbehaving science? First, it helped produce certain intellectual commitments, in particular, the genetic explanation of racial differences, deterministic and "astrological" ways of conceiving genetic effects, and the importance of separating nature from nurture. It reinforced the field's "accumulative" pattern of knowledge production. And it promoted intellectual fragmentation generally and the abandonment of alternative intellectual projects. Provocation flourished as a style of interaction and communication both with other scientific communities and with public audiences. And at the broadest level, the authority and legitimacy of behavior genetics has suffered as has its capacity to secure resources. There has always been a stigma attached to behavior genetics, and it has always been small and modestly supported—especially compared to a field like neuroscience, which, as I showed in chapter 1, was founded under similar conditions yet never suffered the legacy of disruptive controversy.

Behavior genetics is an idiosyncratic field with a distinctive history; as an exemplar of misbehaving science it is extreme. What lessons can we draw about tendencies toward misbehavior in other scientific fields? Would a field have had to experience something as disruptive as the IQ controversy? Is misbehaving science uniquely associated with fields with archipelagic structures?

At its most elemental, the answer, I think, is to look for forces that act centripetally on scientists, drawing them away from mutual engagement. Misbehaving science will be encouraged by forces that de-intensify scientists' competition for mutual recognition, by forces that undermine their investment and attachment to a particular scientific space and encourage easy movement between domains, and by arrangements that push the separation of critical and scientific practices.

This isn't to suggest that "community" among scientists is the antidote to misbehaving science. The idea of community implies that common interests and agreement hold scientists together. Indeed, the situation under which behavior geneticists were acting most like a "community" was when they adopted a bunker mentality, supporting each other without question to fend off outside attacks. At this extreme, community is something like the opposite of science. The crucial issue is not community but productive competition. This, as Pierre Bourdieu and Charles Taylor tell us, is the kind of competition that is only possible because there is deep investment in the pursuit of a set of shared stakes governed by a common set of rules.[8]

Recent work on "undone science" or "agnatology" argues that ignorance, uncertainty, or gaps in knowledge can be actively promoted by social forces.[9] Such studies have emphasized the institutional matrix of public and private funding and political interests as driving forces. Some of the knowledge effects of misbehaving science are similar, but, at least in the case of behavior genetics, they are not driven by crude interference by "extrascientific" forces. Knowledge gaps produced by field-level anomie can be driven by conflicts among scientists (though, of course, there are political and economic stakes at play). Misbehaving science thus suggests potential mechanisms for some kinds of undone science. The institutional matrix surrounding a scientific field could promote disorganization or anomie directly (the next section considers some examples) or it might exert centripetal forces on scientists that de-intensify their mutual scientific competition and the attendant forms of field responsibility and engagement.

Perhaps the most obvious class of these forces involves the elevation of scientists' pursuit of nonscientific forms of capital. Economic and political capital were less operative than we might have expected in behavior genetics, but many fields are quite different. Increasingly, science is becoming an economic engine, and the scientist-entrepreneur is becoming a common type.[10] Scientists are becoming more open about political motivations, and some seek ways to hybridize the pursuit of scientific capital with the symbolic capital of environmental or health activism.[11] Media exposure is becoming a form of symbolic capital that's increasingly fungible across social fields.[12] The issue here is not so much motivations or conflicts of interest within individuals. Weber's image of the singular scientific vocation notwithstanding, people always have a mix of interests and roles. The question is whether fields have centrifugal forces to offset or complement these centripetal ones.

Another, often overlapping, set of forces that would tend to promote misbehaving science is those that are undermining the professional honor systems of science, medicine, and the professoriate more generally. Many transformations in the university may be contributing. Policy makers and administrators increasingly see universities as engines of economic growth, and they have built an array of incentives to direct scientific work in more economically productive directions.[13] Consistent with the idea of misbehaving science, critics have worried about commerce acting as a centripetal force pulling scientists away from professional and academic concerns.[14] Furthermore, in pursuit of prestige and building their "brands," universities are vigorously pursuing academic "stars" and thus rewarding individualism and promoting free agency among the faculty.[15] Just as these changes erode the attachments of academics to particular universities and departments, they may also erode scientific productivity since individualistic stars improve their colleagues' productivity less than more collegial scientists.[16]

Biomedical, especially pharmaceutical, research is another hotspot of professional transformation. Research on new drugs and clinical treatments is increasingly moving out of the academy, being performed by contract research businesses, and being coordinated by publication planners.[17] University researchers are often included more for the symbolic capital of their affiliations than their intellectual contributions. Surely this bureaucratization aids productivity in many ways, given the complexity and coordination problems of biomedical research. However, it also represents a shift in control away from a research system governed by a code of professional norms and recognition toward one of corporate management. These are just a few examples of changes that are serving to disembed scientists from the professional and institutional structures that have long governed the terms of scientific competition and production. It wouldn't be surprising if the seeming rise in public reports of scientific misconduct, false positives and research retractions, conflicts of interest, public mistrust and rejections of scientific authority, and the like is a real trend—toward an era of misbehaving science—reflecting the disorder of a transforming system.

Misbehaving Science and Science Governance

Perhaps it is prudent to consider some of the lessons that this analysis of misbehaving science might hold for two recent trends in the governance of science. The first of these is the widespread enthusiasm for reorganizing science along interdisciplinary lines. Over the last two or three decades it has become popular to assert that interdisciplinarity is a key to revolutionizing knowledge production and innovation and making research more problem driven and accountable to social needs.[18] Interdisciplinary initiatives have been promoted by funders, foundations, and university administrators. The National Science Foundation, in particular, has invested heavily in interdisciplinary science with several high profile programs (for example, INSPIRE, IGERT, and CREATIV) in addition to more routine programs.[19] In polls, substantial majorities of the professoriate also agree that interdisciplinary knowledge is "better than" knowledge emerging from disciplines.[20] If, at its heart, interdisciplinarity is about tightening intellectual connections between disciplines (or epistemic cultures or theoretical and methodological tools), the desire is understandable given the seemingly inevitable tendency toward specialization and fragmentation of scientific practice.

The example of behavior genetics, however, demonstrates empirically that interdisciplinarity is not a homogenous pursuit, nor need it lead to tightened connections. One way to understand behavior genetics' history is that its founders aspired to build an integrated, transdisciplinary science but ended up with a fragmented multidisciplinary space. A shared cognitive commitment to the problem of the genetic inheritance of behavior was not enough to foster integration. Disciplinary barriers persist, and they have often been reinforced by status conflicts among scientists. Further, the kinds of interdisciplinary connections behavior genetics fosters—"giving away" the field's frameworks and tools—produce other kinds of conflict and fragmentation among behavior geneticists who favor other frameworks and between behavior genetics and other disciplines. Intellectual integration is no simple matter, and here, at least, it exists in a dialectical relationship with fragmentation.

The basic motivation for interdisciplinary science is the production of new knowledge. The propagation of behavior genetics tools to new types of scholars certainly creates new scientific opportunities. But as I showed, this had more to do with the accumulation of similar findings in new realms and reproducing a prefabricated argument against environmentalist orthodoxies in the social sciences than making fundamentally new kinds of contributions to knowledge. Is new knowledge always better knowledge? This case suggests it is important to distinguish between interdisciplinary knowledge organized to serve individual scientists' interests in professional production through repetition and that which helps rethink *doxa*.

How well will interdisciplinary fields be set up to manage controversies? Unlike disciplinary knowledge that is supposedly inwardly directed and academic, interdisciplinary knowledge is supposed to be problem based and worldly. This orientation toward problems where many actors have their own stakes and understandings would seem to make interdisciplinary science inherently more inclined to provoke public controversy. Controversy persists in behavior genetics, in part, because of the centripetal forces that loosely bind participants to the field: barriers to entry and exit are low; required investments are few; participants need not seek each other's recognition. Thus there is little to bind behavior geneticists to collective problem solving. The danger for interdisciplinary sciences more broadly is that efforts to dismantle structures and barriers in the name of fostering new intellectual connections ignore the centrifugal forces those structures help sustain. Intellectual connections are the main centrifugal force upon which interdisciplinary science advocates focus. I suspect that these are relatively weak, and without efforts to foster others, interdisciplinary fields are very likely to feature the centripetal forces that produce misbehaving science.

Calls for interdisciplinary science are part of wider enthusiasm for postdisciplinary, translational, and entrepreneurial science—which is to say knowledge production "unbound" by the traditional professional milieu. Advocates (and analysts) of these changes should consider the many possible implications of the kinds of restructuring and de-structuring that would take place. Disciplines tend not to be susceptible to misbehaving science because their structural core, the university-supported production and consumption of graduate students, generates strong centrifugal forces.[21] The question spawned by the problem of misbehaving science is how interdisciplines can overcome the supposed intellectual shortcomings of disciplines without producing the problems of too many centripetal forces. In all these efforts to unbind science from traditional professional structures and cultures, what forces can engender sufficient centrifugal pull?

The case of behavior genetics and the concept of misbehaving science also hold lessons for the tendency for scientific fields to "outsource" ethical and social reflection on the implications of their work. Kelly Moore has documented a long struggle since World War II among scientists over how to cope with the social implications of their research and also how to manage the inevitable politicization of science that such coping entails.[22] The decision of Human Genome Project leaders to devote a substantial budget to funding ELSI projects was a watershed moment. This move was a clear bid to shore up legitimacy given the public controversies surrounding genetics, but it was also a genuine recognition that scientists didn't have the expertise to confront all the issues. More to the point, it helped solidify a model for managing contentious scientific problems. Scientific institutions and funders would outsource consideration of these issues to professional experts, sanctioning and financing their efforts. What does the case of behavior genetics have to say about this general strategy?

In chapter 7, I briefly documented the rise and fall of ELSI projects devoted to behavior genetics. These efforts failed for several reasons. One was their inability to balance the search for a middle-ground consensus with an adversarial structure. Scientific and "ELSI" expertise became pitted against each other and couldn't be reorganized in a different relationship. Different ideas about the boundary between "science" and "implications" couldn't be reconciled. There were other conflicts the sociological literature on bioethics has pointed out.[23] For example, the tendency to identify "implications" with "ethics"—with its focus on rational, moral, and individual-level analysis—posed challenges to other framings and, in particular, made social and cultural analyses look political and emotional in comparison.

But behavior genetics illuminates a much deeper issue: the problems resulting from the division of science from critique. Over the course of their long experience with controversy, behavior geneticists came to see critics as a type whose challenges could be engaged selectively and eventually ignored. Crucially, behavior geneticists understood their critiques—whether technically nitpicky or broadly theoretical—as animated by political concerns and social fears more than science. Behavior geneticists also withheld recognition from their peers who criticized within the field; those who persisted were marginalized. And they created a tacit division between legitimate and illegitimate critical practices. This allowed the technical corrections necessary for peer review to function while censoring expressions of deeper conceptual skepticism.[24] The overall effect? Behavior geneticists essentially outsourced scientific criticism as a way of neutralizing its threat. And many of the field's tendencies toward misbehavior are linked to these arrangements.

"Bunkerization" like that experienced in behavior genetics may be uncommon, but the outsourcing of social and moral analysis of science, the division of scientific from social critique, and unease at publicly criticizing other scientists seem to be rising trends in science. I discussed some of these trends in chapter 7 in describing scientists' reactions to *The Bell Curve*. Sociologist John H. Evans tracked public debate about genetic technologies from 1959 to 1995.[25] Through the early 1980s scientists eagerly engaged the public and were among the most influential voices. After that point their participation and influence dropped off sharply as public debate became dominated by professional bioethicists. Race has become a hot topic in contemporary genomics research. Sociologist Catherine Bliss has shown that, unlike the eugenicists and race researchers of earlier eras, contemporary genomics researchers explicitly claim social justice, the desire to promote equity and inclusion, and the elimination of health and social disparities as motives.[26] But they have not turned this commitment into a critical public voice; they have been mostly unwilling to criticize publicly racial science they consider misguided, largely silent on misuses or miscommunication of genomic science of race in the media, and uneasy about the few vocal scientist and nonscientist critics. Historian Sarah Richardson has echoed this last point in particular: genomics researchers have effectively delegated much of the critical work about race, but have failed to accord any reasonable recognition for critical efforts.[27]

The trend of outsourcing ethical and social reflection on science is thus a dangerous one. It helps reinforce the idea that there is a bright line between "science" and "social implications" and releases scientists from responsibility to consider the latter or the intersection. The outsourcing trend discourages the cultivation of reflexive critical practice as part of scientists' skill set. It also would seem to encourage misbehaving science; the emergence of controversies may become more likely and "bunkerization" a more common management strategy than mutual policing.

Interestingly, behavior genetics might also be a source for imagining an alternative to the outsourcing strategy. A number of behavior geneticists wished members would take the field seriously as an end in itself.[28] They imagined collective venues in which behavior geneticists of different stripes would engage each other seriously. Their hope was grounded in improving the science. Better engagement might reveal unseen connections and roads not traveled. Separate findings might be interpreted in light of each other (such as, why are heritability scores for lab mice generally lower than humans in heterogeneous environments?). Agreement might be reached about the meaning, interpretation, and communication of different methods. How social definitions of traits or populations could affect genetic findings might be analyzed. The point of the vision was to have scientific conversations that were impossible in the current state of affairs, not to make behavior geneticists more "responsible" or to address "implications." But the scientific issues raised are all *entangled* with definitions and conceptions of the social; such an engagement would reframe the science/implications relationship.

While the issues in play would be field specific, the idea that scientists might engage each other scientifically in ways that question the entanglement of social conceptualizations might have broad implications. The implication here is not that ethical, legal, or social expertise has no role. Rather, this expertise might be put in a different relationship to science if scientists engaged different practices of thinking about their own practice. Specifically, the ELSI burden of representing "the social" in opposition to science might be lifted or shifted. Cultivating this kind of lasting scientific engagement would not be easy; current arrangements of ethical outsourcing seem to be a clear barrier.

Knowledge in Shackles

Early in this book, I aligned my project with Craig Calhoun's tradition of critical theory.[29] I have already described and exemplified the methodological, or better yet, analytical dimension of this critical project. I could not have analyzed behavior genetics or misbehaving science without breaking with previous approaches and the categories they used; field theory was a core tool of this effort. What critique of behavior genetics has this analysis made possible?

Others have criticized behavior genetics epistemically, arguing that it has not or cannot support the claims it makes, or criticized its implications, arguing that the science misleadingly bolsters dangerous political and social ideas.[30] My critique is based on the problematic knowledge and knowledge production that flows from the misbehaving field. First, what actually counts as the science of "behavior genetics" is much more limited than what might have been possible—it is an intellectual space littered with gaps, missed connections, and "undone science."[31] Behavior genetics has come to be largely identified with analyses of genetic contributions to behavioral variance (using quantitative or molecular methods) in humans. The label doesn't apply to many other articulations of the gene/behavior problem. More than just a label is at stake because the fragmentation of the field means that there is limited engagement and conflict with other scientific perspectives that might have been mutually informative or complementary if they were under the same tent. Beyond the fragmentation of existing perspectives, I have several times pointed to the foreclosure of various intellectual roads in the course of the field's development. Without speculating about the merits and problems of an alternate history, we can assert that misbehaving science has restricted what might be known about genes and behavior.

If the restricted breadth of behavior genetics is one point of criticism, a second is its characteristic fixations. First among these is its defense of the genetic hypothesis for racial differences. For more than forty years, behavior genetics has been the uneasy launching pad for ill-founded and socially pernicious claims about racial differences. As I have shown, this is due not to any unique scientific insight but also to the field's misbehaving characteristics. The second is the field's commitment to the idea that the way to think about the gene/behavior relationship is through the nature/nurture framework. Whatever the version—old-fashioned nature *versus* nurture, nature *via* nurture, or the interactionist nature/nurture *interplay*—they are all predicated on conceiving the two as ontologically distinct causes and believing the goal of science to be separating and measuring them.[32] Some have argued that this is the very essence of behavior genetics.[33] My analysis shows otherwise. At the field's founding behavior geneticists declared the nature versus nurture debate over and were eager to pursue alternatives. Instead, they spent the next fifty years litigating the controversy. This was not a worthwhile expenditure of scientific energy. Not only has behavior genetics put the nature/nurture framework at the center of its concerns, it has also reinforced an extreme interpretation. In broad strokes, this interpretation understands the framework deterministically, or at least causally; "astrologically" as explaining individuals' fates and differences; and it reifies the analysis of variance of a behavior as the explanation of the behavior itself.

New developments may be walking science down some of those untraveled roads. Analysts of contemporary genetics have argued that postgenomic developments have systematically dismantled the nature/nurture framework.[34] Maybe so, but its death knell has been sounded many times before. Whatever happens, the legacy of misbehaving science will be felt for a long time. Claims and arguments will be made in an archipelagic space where the rewards for mutual engagement and critique are few. It is telling that social science critics of behavior genetics have advanced these charges, not research scientists. This disorderly scientific space continues to be one where it is easy to lob bombs at orthodoxies. But it seems much will have to change socially before it is a place where *doxa*—unthought taken-for-granted assumptions—can be systematically dismantled and constructive alternatives reassembled.

This critique of science does not target the actions of individuals or even the properties of an episteme. It focuses on a structural logic and whether it closes or opens possibilities for knowledge. We might think that intellectual and social de-structuring of science, the loosening of restrictions and policing by ones peers, would be scientifically freeing. Unencumbered by rules, expectations, and criticism, shouldn't intellectual creativity flourish and horizons expand? The basic lesson of behavior genetics and misbehaving science is that science liberated from structure leads to knowledge in shackles. Just as we can criticize behavior genetics in terms of the possible not the actual, our sociology can be oriented toward understanding the features of science that open, rather than constrain, possibilities for knowledge.

Appendix

Interviewees' home disciplines and positions with respect to behavior genetics

Interviewee #	Discipline	Field position
1	Biology	Critic
2	Human genetics/Epidemiology	Psychological behavior genetics
3	Neuroscience	Animal behavior genetics
4	Molecular genetics	Critic
5	Psychology	Psychological behavior genetics
6	Psychiatry	Psychiatric genetics
7	Neuroscience/Psychiatry	Animal behavior genetics
8	Genetics	Psychological behavior genetics
9	Sociology	Critic
10	Psychology/Psychiatry	Psychiatric genetics
11	Genetics	Animal behavior genetics
12	Genetics	Animal behavior genetics
13	Molecular genetics	Molecular genetics
14	Psychology	Psychological behavior genetics
15	Behavioral epidemiology	Psychological behavior genetics
16	Psychology	Animal behavior genetics
17	Sociology	Commentator
18	Psychology	Critic
19	Population genetics	Critic
20	Psychology	Psychological behavior genetics
21	Psychology	Psychological behavior genetics
22	Biopsychology	Animal behavior genetics
23	Psychology	Psychological behavior genetics
24	Psychiatry	Psychiatric genetics
25	Psychology	Psychological behavior genetics
26	Philosophy	Commentator
27	History/Political science	Critic
28	Psychology	Psychological behavior genetics
29	Psychology	Psychological behavior genetics
30	Neuroscience	Critic
31	Psychiatry	Psychiatric genetics

32	Psychology	Psychological behavior genetics
33	Pharmacology	Molecular genetics
34	Molecular genetics	Animal behavior genetics
35	Psychology	Psychological behavior genetics
36	Psychology	Animal behavior genetics

Notes

Introduction

1. Herrnstein and Murray (1994).

2. Rushton (1994).

3. Rushton (1994, esp. 199–216).

4. Whitney (1995b). Winston and Peters (2000) showed that Whitney had dishonestly or at least carelessly constructed his data to produce this claim. See also Whitney (2000).

5. Whitney (1995b). The personalization isn't apparent in the printed version of the speech, but this is according to people who were there.

6. Interviewee #7. See also Butler (1995); Holden (1995a; 1995b).

7. Naureckas (1995).

8. See Rushton (2000); see also Horowitz (2000).

9. For a collection of relevant documents, see Jacoby and Glauberman (1995).

10. Butler (1995); Holden (1995a; 1995b).

11. See Brimelow (1994) and various selections in Jacoby and Glauberman (1995).

12. Fraser (1995); Kincheloe, Steinberg, and Gresson (1996); A. Miller (1995); Sedgwick (1995).

13. Devlin et al. (1997); Fischer et al. (1996).

14. Botstein (1997, 212).

15. Hirsch (1997a).

16. Wahlsten (1995).

17. Arvey and al. (1994); see also Gottfredson (1997). Fifteen of the signatories were members of the Behavior Genetics Association in 1996 (BGA [1996]). Ten others have had some degree of engagement with the field of which I am aware.

18. Arvey et al. (1994).

19. Interviewees #14 and #35.

20. Interviewees #14 and #23. See also Horowitz (1995).

21. Interviewees #7 and #36 and Holden (1995a). See also Allen (1998).

22. Duke (1998).

23. Interviewee #28. There is some dispute about whether molecular genetics will eventually allow a possible finding of racial genetic differences (for example, interviewees #8 and #13). However, as this speaker points out, blacks and whites face different environments, so if they also have different genes associated with intelligence, the causal ambiguity still will not be cleared up.

24. Interviewee #23.

25. Friedrichs (1973); Harwood (1977); Lieberman (1968); Lieberman and Reynolds (1978); Pastore (1949); Sherwood and Nataupsky (1968).

26. Gould (1996); Kamin (1974).

27. Lombardo (2002); Mehler (1997); Mehler and Hurt (1998); Miller (1994); Tucker (1994; 2001; 2002).

28. Lewontin, Rose, and Kamin (1984).

29. Interviewees #10, #15, #35, and #36. See also Harwood (1977); Lewontin (1976c); Lykken (2002); Scarr (1987), who describe some behavior geneticists as politically liberal. Even if numerically there are a lot of conservatives, it would be necessary to explain how they are able to impose their views on the liberals under the guise of science.

30. Gottfredson (1994) has also accused behavior geneticists of having liberal cowardice on this issue. See also Nyborg (2003); Whitney (1995a).

31. In Holden (1995b).

32. Interviewee #8. Conservative commentator Jim Manzi (2008) has argued that the impulse to intervene and correct, which animates behavior genetics and related sciences, is precisely the reason those on the right should be wary of the field.

33. Lakatos (1970); Popper (1962).

34. Collins (1985); Shapin (1994).

35. Eysenck (1971); Jensen (1969); Shockley (1992 [1968]).

36. Evans et al. (2005); Mekel-Bobrov et al. (2005); Richardson (2011).

37. Rowe (1994); Scarr (1992).

38. Baumrind (1993); Jackson (1993); Maccoby (2000).

39. See Hamer and Copeland (1998) and Kaplan (2000) for supportive and critical reviews respectively.

40. Allen (1999); Farahany and Bernet (2006); Feresin (2009).

41. Hamer and Copeland (1994); Hamer et al. (1993).

42. Conrad (1997); D. Miller (1995).

43. Parens (2004); Parens, Chapman, and Press (2006).

44. Collins (1981; 1985; 2000); Gieryn (1999); Jasanoff (2004); Latour (1987).

45. Bourdieu (1975); Collins (1985; 2000); Gieryn (1999); Shapin (1994); Simon (2002).

46. The kinds of norms relevant to scientific misconduct and misbehaving science are often at different levels. As is the case in behavior genetics, it is possible to have a functioning peer review apparatus—and thus the effective policing of certain scientific standards—while other kinds of norms are absent or unclear. As many sociologists of science have pointed out, scientists' mutual skepticism and criticism are always constrained relative to the doubts they might express rationally (Bourdieu [1975]; Collins [1985]; Shapin [1994]). But in the kind of misbehaving science behavior genetics exhibits, peer review is particularly constrained in its reach or effectiveness. In behavior genetics it tends to become focused on a narrow range of technical matters while deeper critical questions are not raised. Indeed, as some of my interviewees have reported, peer review is a site where criticism and questioning are criticized and censored. Paradoxically, in behavior genetics normlessness is propagated by norms against asserting norms.

47. Mitroff (1974); Mulkay (1976).

48. Celebratory examples include Hamer and Copeland (1998) and Wright (1998); critical examples include Kaplan (2000) and Moore (2001); and those focused on its implications include Fukuyama (2002) and Nelkin and Lindee (2000). The next chapter considers different approaches to analyzing behavior genetics in more detail.

49. Bourdieu (1975).

50. My selection criteria ensured representation from all the field's important quarters, but it "oversamples" on the field's eminent and long-term members because they were most likely to have experience and knowledge with the field's history and controversies. Overall I contacted fifty-five individuals by e-mail, and of those with whom I was unable to conduct interviews, six did not reply to my requests, nine declined to be interviewed, and five agreed initially but the interview was not conducted for scheduling reasons, because we could not come to terms on interview arrangements, or because my subsequent efforts to make contact were not heeded. Most of those who declined to be interviewed cited lack of familiarity with the field or time constraints.

At various points in the book I attempt to interpret the significance of some nonresponses. But since all the "critics" I contacted agreed to be interviewed and at least two behavior geneticists declined because they perceived my project to be dangerous to themselves or the field, it is clear (and not surprising given the field's fraught relationship with sociology) that my effort was perceived probably to be critical in intent. Given this situation the response rate and generosity from interviewees is remarkable. If this produces a possible bias, it is likely that my interviewees represent relatively "mainstream" perspectives rather than those who are likely to be more suspicious of sociology or efforts to analyze the field from the outside. However, many of these efforts are represented in print, so I did not lose them completely.

My interview schedule was semistructured. I asked interviewees about their backgrounds and how they entered the field; their sense of the field's history, its scientific terrain, and where their work fit in; and their knowledge of, experience with, and opinions of controversy in the field. The interviews, lasting from thirty minutes to three hours, were recorded and transcribed. In quotations, I use ellipses (. . .) to indicate excisions and long dashes (—) for pauses or breaks in the dialogue. Most interviews occurred face to face, but I conducted five by telephone and three respondents preferred dialogue by e-mail.

Because the field is small and members well known to one another, respondents were interviewed without the promise of anonymity (but they could make statements off the record). Usually I do not identify interviewees by name, except in the few cases where it would not make sense to do so. See the appendix for the list of interviewees by subfield.

51. Ziman (2000).

Chapter One

1. Bourdieu (1975; 1988; 1991; 2004). See also Bourdieu (1993; 1996). Bourdieu has been a major inspiration on the rise of field theory in analyses of the intersection of political and intellectual fields; for example, Eyal (2002); Medvetz (2012); Stampnitzky (2013); and also Fligstein and McAdam (2012); Martin (2003). Despite ambivalence and criticism (e.g., Curtis [2003]; Mialet [2003]; Sismondo [2011]), Bourdieu's field theory is also attracting interest in the sociology of science; for example, Albert, Laberge, and Hodges (2009); Kim (2009); and the articles in a special issue of *Minerva*, "Beyond the Canon: Pierre Bourdieu and Science and Technology Studies" (49 no. 3 [2011], http://link.springer.com/journal/11024/49/3/page/1).

2. Bourdieu (1975, 19).

3. Bourdieu (1990; 1998).

4. Bourdieu (1998, 76-77).

5. For a sample of logical analyses, see Griffiths and Tabery (2008); Kaplan (2000); Moore (2001); Rutter (2006); Tabery (2009); Urbach (1974a; 1974b). For some emphasizing political motives, see Allen (1996; 1999); Alper and Beckwith (1994); Friedrichs (1973); Gould (1996); Harwood (1977); Hirsch (1975; 1981); Hunt (1999); Nyborg (2003); Paul (1994; 1998c); Sherwood and Nataupsky (1968); Tucker (1994).

6. Gottesman (2008); Loehlin (2009); Maxson (2007).

7. Griffiths and Tabery (2008); Pastore (1949); Scarr (1987); Tabery (2009).

8. Diane Paul (1994) makes a similar point about how the changing intellectual terrain made geneticist Theodosius Dobzhansky, once considered an environmentalist, later appear hereditarian.

9. Ari Adut (2004; 2008) has defined scandal as the "disruptive publicity of transgression." His analysis of scandal as linked to activities of norm entrepreneurship has informed much of my thinking on controversy in behavior genetics.

10. Abbott (1988).

11. Panofsky (2011).

12. My extension of field theory to a heterogeneous and poorly bounded social space is far from unique. Indeed, recent books by Thomas Medvetz (2012) on think tanks and Lisa Stampnitsky (2013) on terrorism experts have shown field theory to be well adapted to explaining the distinctive social dynamics of ambivalently structured spaces that are situated interstitially between more established fields.[F]

13. Durkheim (1951).

14. Alexander (1995).

15. Taylor (1985; 1993).

16. These forces are not the same as members of a community who all carry a common set of norms in their heads, as Durkheimian integration is often interpreted. Nor are they Merton's institutional imperatives, disconnected from practices or conflict. Fields of forces were Bourdieu's way of retheorizing the conditions of integration necessary for agonistic action without positing common consciousness, shared identity, or emergent institutions separated from the actions that create them.

17. Clarke (1998); Collins (1985); Fujimura (1996); Knorr-Cetina (1999); Latour and Woolgar (1986); Lynch (1985); Shapin (1994).

18. Philosopher Jonathan Kaplan (2000) analyzes behavior genetics controversies topic by topic.

19. Giddens (1984).

20. Nicole Nelson's (2011) study of an animal behavior genetics laboratory is an example of the local, ethnographic approach.

21. Cambrosio and Keating (1983); Collins (1998); Frickel (2004a); Frickel and Gross (2005); Lemaine et al. (1976); Mody (2011); Whitley (2000).

22. Frickel (2004b) and Shostak (2013) similarly emphasize the developmental effects of structural weakness in their case studies of genetic toxicology and public health genetics respectively.

23. Bloor (1991).

24. Alper and Beckwith (1994); Duster (2006); Fukuyama (2002); Habermas (2003); Kaplan (2006; 2000); Nelkin (1999); Nelkin and Lindee (2000); Panofsky (2009); Parens, Chapman, and Press (2006); Rowe (1997).

25. For authors that trace these links and breaks, see Allen (1996; 1999); Duster (1990); Gould (1996); Lewontin, Rose, and Kamin (1984); Loehlin (2009); Maxson (2007); Mehler (1997); Paul (1998c); Rafter (1997); Tucker (1994).

26. Calhoun (1993, 63).

27. Hacking (1999).

28. Bourdieu (2004); Bourdieu and Wacquant (1992).

29. Interviewees #2, #5, #13, #15, and #23.

30. Parens (2004).

31. Tabery (2009). See also Griffiths and Tabery (2008).

32. Kevles (1985).

33. Ramsden (2002).

34. Provine (1986). See also Barkan (1993).

35. Indeed, as I will show in chapters 3 and 4, the fields of genetics and behavior genetics substantially parted ways in the mid-1970s over just this issue. Since that point, behavior genetics has been much more associated with psychology and psychiatry than genetics proper.

Also, Richardson (2011) has suggested that the increasingly hybrid and transdisciplinary organization of contemporary genetics is undermining the will and capacity to engage in critiques of racial science. This argument fits well with the theme of misbehaving science.

36. See Wright (1986). Indeed, critics have argued that geneticists were less interested in fundamentally addressing safety than setting the stage for the emergence of a minimally regulated biotechnology industry.

37. On animal research protests, see Jasper and Nelkin (1992) and Society for Neuroscience (2003). On reductionism, see Spear (2000).

38. Oreskes (2004).

39. Oreskes and Conway (2010).

40. http://en.wikipedia.org/wiki/Intelligent_design_in_politics, accessed April 6, 2012.

41. Nelkin (1984).

42. http://en.wikipedia.org/wiki/Climatic_Research_Unit_email_controversy, accessed April 17, 2012.

43. Goldenberg (2012).

44. Goldenberg (2012).

45. Revkin (2012).

46. http://www.ipcc.ch/organization/organization_procedures.shtml and http://www.ipccfacts.org/how.html, both accessed April 17, 2012.

47. Nestle (2002).

48. Pollan (2009); Scrinis (2008); Taubes (2007).

49. Nestle (2002). Also, Stephen Hilgartner (2000) shows how nutrition scientists debated each other about the virtues of offering population-wide nutrition advice. The concern was that in engaging the unruly, heteronomous space of diet

advice, they would lose their authoritative, supposedly separate status and become just another voice.

50. Chimonas et al. (2011).

51. Mirowski and Van Horn (2005); Sismondo (2009).

52. Ioannidis (2005); Naik (2011); Zimmer (2012).

53. Weisz et al. (2007).

54. Blume (2006); Hess (2004); Moore (2008).

55. On neuroscience, its scope and structure, see Spear (2000). Neuroscience contrasts with specialties that form to exploit a narrow intellectual niche, technological innovation, or a particular set of questions. See examples in Lemaine et al. (1976).

56. Diane Paul (1998b) has argued that much of medical genetics *was* behavior genetics in its early years, so the separation of the two fields—medical genetics' growing unease with behavioral traits—is part of the story.

57. See Panofsky (2011).

58. See http://en.wikipedia.org/wiki/Neuroscience, accessed April 17, 2012.

59. Anderson (1983).

60. In 2011–12 the membership of the Behavior Genetics Association was 444 (down from 578 in 1996), International Society of Psychiatric Genetics was 550, and the International Behavioural and Neural Genetics Society was 228. Adding them together yields 1,222, but this is an overestimate because some people are members of multiple societies.

61. http://www.sfn.org/index.aspx?pagename=membership_AboutMembership, accessed April 17, 2012.

62. For lists of relevant journals, see http://www.ibangs.org/links and http://en.wikipedia.org/wiki/Behavior_genetics, both accessed April 17, 2012.

63. Spear (2000, 118).

64. Spear (2000, 109–20). See figures 6.1 and 6.2 for behavior genetics' growth in terms of grants and funding.

65. Turner (2000).

66. Spear (2000, 119).

67. Some of these include the Institute for Behavior Genetics at University of Colorado, the Social, Genetic & Developmental Psychiatry Research Centre at King's College London, and Virginia Institute for Psychiatric and Behavioral Genetics at Virginia Commonwealth University, plus the departments of psychology at Minnesota, Texas, and Penn State and psychiatry at Washington University. See table 4.2 for more details.

68. The behavior genetics figure comes from a search of a website (www.sunshine-project.org/crisper/) that gives funding amounts for the NIH grant database (accessed August 20, 2007, site no longer exists). Because the keywords *behavioral genetics* and *neuroscience* are not used in the same way, the website cannot give comparable figures for the two fields. Baughman et al. (2006)

report that a $25 million initiative is 0.6 percent of the annual NIH budget for neuroscience.

69. In 2003, 1998, and 1993, 6 percent–9 percent of articles published in *Behavior Genetics* and 22 percent–33 percent of those in *American Journal of Human Genetics* (*Neuropsychiatric Genetics*) cited private funding sources in the acknowledgments.

70. See the Society for Neuroscience website: www.sfn.org, accessed September 10, 2007.

71. Lombardo (2002); Lynn (2001); Tucker (2002).

72. Wilayto (1997).

73. Segal (2012, 309–19, 329–31). Scientists organizing MISTRA, which has not made claims about racial differences, apparently debated taking Pioneer money, but also insist that the fund has not tried to influence research.

74. I searched the acknowledgments of articles published in *Behavior Genetics* in 1973, 1978, 1983, 1988, 1993, 1998, and 2003. The only evidence I found was the Pioneer Fund listed twice, and both were for presentation abstracts (one each by Jensen and MISTRA). Hirsch (1997b) reports research funding for 1990–91 and 1992–93 for the field's leading research institute, the Institute for Behavioral Genetics at the University of Colorado. Of the $9.3 million received for the two years, 87 percent was from federal sources. About $300,000 over two years came from R. J. Reynolds, the tobacco corporation, and about $57,000 over one year from the free-enterprise promoting Adolph Coors Foundation. The MacArthur Foundation was the other private funder listed. Funders who might have a political or economic interest in the outcome of research were a small source of money.

75. Allen (1999); Fishbein (1996); Green (1985); Rose (2000).

76. Conrad and Weinberg (1996); Lite (2009); Macrae (2009); Mello (2007); Muir (2001); Wade (2009).

77. Nelkin and Lindee (2000).

78. Alper and Beckwith (1994); Kaplan (2000); Panofsky (2009); Rowe (1997).

79. Rowe (1997); see also Kaplan (2000).

80. Rowe (1997).

81. Harwood (1979); Nelkin (1999).

82. Carson (2007).

83. Epstein (1996).

Chapter Two

1. For histories of eugenics, see Carlson (2001); Kevles (1985); Ludmerer (1972); Paul (1998a; 1998c); Ramsden (2008).

2. Barkan (1993); Kevles (1985); Provine (1986).

3. Carson (2004; 2007).

4. Carson (2004, 202).

5. Fuller and Thompson (1960, 1).

6. Parsons (1967, 5, emphasis added).

7. For example, Gould (1996); Lewontin, Rose, and Kamin (1984).

8. Griffiths and Tabery (2008); Harwood (1976); Moore (2001).

9. Dewsbury (2009); Paul (1998d).

10. Paul (1998d).

11. Scott (1985) and Dewsbury (2009).

12. Scott (1947).

13. Dewsbury (2009).

14. Scott and Fuller (1965); see also Paul (1998d). These findings also undermined the foundation's goal of breeding the perfect pet to sell to the public.

15. Scott (1985, 404).

16. Hall (1951).

17. This issue is considered further in chapters 4 and 5.

18. Fuller and Thompson (1960, vi).

19. Fuller (1985, 109).

20. Bliss (1962, ix).

21. Hirsch (1967d).

22. Ramsden (2008); also Kevles (1985, 171–73, 252).

23. Osborne and Osborne (1999c).

24. Osborne and Osborne (1999a) describe these events. Indeed, helping to assemble the behavior genetics network and agenda was among the AES's last-gasp efforts. In 1969 *Eugenics Quarterly* was renamed *Social Biology*, and in 1973 AES changed its name to the Society for the Study of Social Biology. In this period it lost the resources to do much more than fund the journal (Osborne and Osborne [1999b]).

25. These patterns can be seen in the rosters of participants of edited collections that began as conferences during this period. See, for example, Hirsch (1967a); Osborne and Osborne (1999a); Scott (1947); Spuhler (1967); Vandenberg (1965).

26. Vandenberg and DeFries (1970).

27. Osborne and Osborne (1999c).

28. For example, Griffiths and Tabery (2008); Harwood (1976).

29. For information on these figures' ideas, see Kevles (1985); Provine (1986); Tucker (1994).

30. Scott (1985, 411). On Yerkes, see Gould (1996, 222–63).

31. The volume was Vandenberg (1965). Hirsch (1967a, xvi) attributes this conference to Cattell's suggestion. Cattell's biography and reactionary political views have been discussed deeply in Mehler (1997).

32. Osborne and Osborne (1999a).

33. SSRC archives at the Rockefeller Foundation Archives; Accession 2; Series 1 Committee Projects; Subseries 39 Committee on General Correspondence; Box 188; Folder 2148—Genetics and Behavior 1960–1961.

34. Kevles (1985, 207). Kallmann died in 1965, just a few years later. http://en.wikipedia.org/wiki/Franz_Josef_Kallmann, accessed May 9, 2012.

35. Paul (1998d, 74). See also Lindee (2005, 120–55).

36. Beatty (1994); Dobzhansky (1973b).

37. Osborne and Osborne (1999a). The Behavior Genetics Association would later name its distinguished career award the Dobzhansky Prize.

38. See Glass (1968, 178–226).

39. Gottesman (1965, 73).

40. Dobzhansky (1962); Dunn and Dobzhansky (1952). See also Beatty (1994).

41. Fuller and Thompson (1960, 327). Although preventing "the conception of children likely to be severe social burdens," presumably due to extreme genetic disease, was within the purview of their recommendations.

42. See Higgins, Reed, and Reed (1962) and Bajema (1968).

43. Ramsden (2008).

44. Hirsch (1967b, 128). Incidentally, this is precisely the argument that Richard Herrnstein would make publicly and provocatively just a few years later during the IQ controversy (Herrnstein [1971; 1973]).

45. Hirsch (1967b, 129).

46. Gottesman and Erlenmeyer-Kimling (1971, S1).

47. Gottesman and Erlenmeyer-Kimling (1971, S2).

48. Gottesman and Erlenmeyer-Kimling (1971, S4, S6).

49. For example, Dobzhansky (1968; 1973a); Dunn and Dobzhansky (1952). See also Paul (1994).

50. Dunn and Dobzhansky (1952, 134).

51. Fuller and Thompson (1960, 324).

52. Dunn and Dobzhansky (1952, 132).

53. Hirsch (1968, 43).

54. Hirsch (1963); Spuhler and Lindzey (1967). See also Boyd (1950); Lewontin (1972).

55. Hirsch (1962; 1963; 1968).

56. Fuller and Thompson (1960, 323–24). See also Scott and Fuller (1965, 4) and Spuhler and Lindzey (1967, 406).

57. Cooper and Zubek (1958); see also Kaplan (2000).

58. Scott and Fuller (1965) cited in Paul (1998d, 74–75).

59. Fuller and Thompson (1960, 324).

60. Spuhler and Lindzey (1967, 366).

61. Spuhler and Lindzey (1967, 384, 388).

62. Spuhler and Lindzey (1967, 406, 408–9). They also reviewed the PKU evidence. This is a single gene that produces a disorder that features mental retardation and thus lower intelligence. It was believed differentially distributed by race (there were no Jewish or Negro cases in a major study), though this would not account for population distributions of intelligence.

63. Spuhler and Lindzey (1967, 414).

64. Spuhler and Lindzey (1967, 405, 413).

65. Spuhler and Lindzey (1967, 414).

66. In another venue Lindzey commented: "reluctance [to publish on behavioral correlates of eye color] in this area is the obvious reluctance in this country and perhaps in any democratic society, to report data which can be used by race fanatics. . . . There is very little [research in the area of racial or Jewish/non-Jewish intelligence differences]; and it is not reported for the same good democratic, but bad scientific values" (Vandenberg [1965], 333–34). See also Hirsch (1968).

67. Carlson (2001); Kevles (1985); Ludmerer (1972).

68. In a conversation transcribed in Vandenberg (1965, 334), when Lindzey says that race differences are not reported for political rather than scientific reasons, psychologist Daniel Freedman then says it might not be reported because the science is poor. Lindzey retorts that with the number of psychological journals, even data of dubious quality could be published somewhere.

69. Fuller and Thompson (1960, vi).

70. Scott (1985, 411).

71. Fuller and Thompson (1960, vi).

72. Bliss (1962, ix).

73. Fuller and Thompson (1960, 95).

74. Hirsch (1967b, 126)

75. Hirsch (1967b, 128)

76. Hirsch (1967b, 127).

77. Vandenberg (1965, 310). In the 1967 Princeton Conference, it seems that a version of this conversation was reiterated. Although here Hirsch was on the biological mechanism side and Scott and Ginsburg thought that things like motive and emotion should be considered. That is, they thought focus should be on behavior not mechanism.

78. Vandenberg (1965, 308).

79. Vandenberg (1965, 312).

80. Vandenberg (1965, 304–15). The suggestion that statistical factors (such as g or general intelligence) existed because of some kind of biological substratum would become a major bone of contention during the IQ and race controversy considered in the next chapter.

81. Skinner (1953); Watson (1930).

82. Watson (1930, 82).

83. Scarr (1981, x).

84. Hirsch (1967c; 1970).

85. Hirsch (1963).

86. Hirsch (1968). Osborne and Osborne (1999c, 244) describe a conflict over this issue in the second Princeton conference in 1965 between geneticists interested in variation and psychologists interested in comparing group means.

87. Hirsch (1963, 1441).

88. Hirsch (1963, 1441–42).

89. Hirsch (1963).

90. Hirsch (1968, 43).

91. Hirsch (1968, 44).

92. Hirsch (1968, 43).

93. Vandenberg (1965, 311).

94. Hirsch (1967c, 426). To be extra clear on the target of explanation Hirsch wanted to rename the field "behavior-genetic analysis" rather than behavior genetics.

95. Hall (1951, 327).

96. Anastasi (1958, 197).

97. Fuller (1960, 43).

98. Anastasi (1958, 197).

99. These included twin registries developed by the National Research Council of veterans; the National Merit Scholarship of high school students; and other state, university, and international sources. See Lindee (2005); Loehlin and Nichols (1976); Vandenberg (1965, 322–29).

100. Fuller and Thompson (1960) quoted in Hirsch (1967c, 420).

101. Hirsch (1967c, 422).

102. Fuller (1960); Fuller and Thompson (1960).

103. Hirsch (1967c, 423).

104. Parsons (1967, 154).

105. Freedman (1968, 1).

106. Bernhard (1967, 32).

107. Paul (1998d, 64–65) argues that much medical genetics research originally was concerned with behavioral issues like intelligence, mental defect, schizophrenia. This connection was based on the idea that lots of disease has an emotional component and also that counseling would be an important medical genetics function, so mental traits would matter. The idea that medical genetics should only focus on specific physical defects was a later development. Early on, medical genetics was deeply entangled with eugenic concerns about human quality; the idea of the "total genetic potential" (mental and physical) was a matter of concern (see also Paul [1998b]).

108. See the agenda of the Third Princeton Conference (Osborne and Osborne [1999c, 245] and *Eugenics Quarterly* 14 [1967]: 199–237). See also Spuhler (1967) and Lindee (2005).

109. Anastasi (1958).
110. Similar ideas also appear in several chapters of Hirsch (1967a, 107–210).
111. Rosenthal (1968, 75).
112. Ginsburg and Laughlin (1966, 247).
113. Fox (1976); Freedman (1968).
114. See Glass (1968, 79–128); also Scott (1976).
115. Rosenthal (1968, 78).
116. Paul (1998d) describes explicit resistance to such fatalism.
117. For example, Bliss (1962); Glass (1968); Hirsch (1967a); Spuhler (1967); Vandenberg (1965).
118. For example, Scott (1976).
119. Ginsburg and Laughlin (1966); also Spuhler (1967).

Chapter Three

1. Jensen (1969).
2. Ezrahi (1976): Harwood (1976; 1977); Snyderman and Rothman (1988); Tizard (1976); Tucker (1994); Urbach (1974a; 1974b).
3. UNESCO (1969); also Provine (1986).
4. Boas (1940); Dunn and Dobzhansky (1952); Klineberg (1963); Montagu (1964). A small sample of the efforts: Anthropologist Franz Boas studied human morphology and found that the children of immigrants, far from being fixed racial types, came to fit American averages for size and stature. Psychologist Otto Klineberg's research showed that black children leaving the segregated South achieved gains in education and intelligence. Geneticists L. C. Dunn and Theodosius Dobzhansky explained the population-based style of modern genetic thought that refused the notion of distinct racial types and argued that rather than fixed traits, the capacity for adaptability was what individuals inherited genetically. Also, Lewontin (1996) notes that egalitarianism was not the norm in genetics or science broadly, but the midcentury aberration. Thus the period between 1950 (UNESCO) and 1969 (Jensen) was the exception despite the fact that race scientists bitterly complain about egalitarian orthodoxies (Gottfredson [1994]; Rushton [1998]; Whitney [1995a]).
5. Barkan (1993); Provine (1986); Reardon (2005).
6. Tucker (1994, 142–44).
7. Harwood (1979); Tucker (1994).
8. Tucker (1994). Psychologists Henry Garrett, Frank McGurk, and R. Travis Osborne claimed that blacks were inherently less educable than whites and that other psychologists were judging the data ideologically. Dwight Ingle discussed the physiological dangers of miscegenation and used his membership in the National Academy of Sciences and editorship of *Perspectives in Biology and Medi-*

cine to provide an outlet for others. Physiologist Wesley Critz George's *Biology of the Race Problem*, commissioned by the governor of Alabama, argued that studies of anatomy and evolution demonstrated blacks' "primitive" nature.

9. Shockley (1992); Shurkin (2008).

10. Shockley (1992 [1974]).

11. Hirsch (1981, 6) offers an extremely detailed account of Shockley's campaign that only a participant in the field could provide.

12. Council of the Academy (1968).

13. Ad Hoc Committee on Genetic Factors in Human Performance (1972); see also Shockley (1969). Lysenko was a Soviet agricultural scientist who held key leadership positions in Soviet genetics from the late 1920s to the 1960s with the support of the Communist Party. Lysenko purged scientists who espoused the Mendelian theory of genetics and thus was a symbol for Western scientists of political suppression of science and genetics in particular.

14. Hirsch (1967c, 421, 434) briefly mentions Shockley.

15. As a postdoc working under Hans Eysenck at the University of London he had been exposed to hereditarian psychology but hadn't published in that mode. Miele (2002, 11–12).

16. Jensen (1969, 82).

17. Snyderman and Rothman (1988).

18. For example, Goldberger (1976); Hirsch (1975); Jensen (1972); Scarr-Salapatek (1976).

19. Davis (1975); Jensen (1972); Rushton (1998); Scarr (1987); Wilson (1978).

20. Eysenck (1971); Herrnstein (1971; 1973); Shockley (1992); Vernon et al. (1970); on Cattell see Mehler (1997).

21. On the broader issues the IQ controversy raised, see Block and Dworkin (1974); Mensh and Mensh (1991).

22. Beckwith (2002); Goldberger (1979); Gould (1996); Jencks (1980); Layzer (1974); Moore (2008); Rose, Hambley, and Harwood (1973).

23. Interviewee #11, also interviewees #12, #14, #28, #36.

24. Hirsch (1970).

25. Hirsch (1975).

26. Hirsch (1975; 1981); Hirsch, Beeman, and Tully (1980a); Hirsch, McGuire, and Vetta (1980b); Hirsch and Tully (1982).

27. Kamin's talk to the ESA was actually in 1973 according to Hirsch (1975).

28. Interviewee #28.

29. See Gottesman (1974, 588). I have taken these quotes slightly out of context. Psychological behavior geneticist Irving Gottesman attributed the virtue of the middle ground to Dobzhansky's book *Genetic Diversity and Human Equality* (Dobzhansky [1973b]). It is clear, however, that Gottesman hoped behavior genetics could emulate this positioning.

30. Ehrman, Omenn, and Caspari (1972). Confronting this controversy was seen as a wide responsibility of behavior geneticists, including animal researchers and others who didn't have a *direct* intellectual stake in the ideas but did have a broader stake in the field.

31. DeFries (1972).

32. See Goldberger (1979) for a comparative analysis of the major statistical models in play at the time.

33. Scarr and Weinberg (1976).

34. Scarr et al. (1977).

35. Nichols and Anderson (1973).

36. Scarr (1981).

37. Scarr (1981; 1987).

38. Jensen (1969) had actually tried to enroll animal research into his argument as well. However, apart from Hirsch (1975), who attacked Jensen's representation of mouse breeding experiments, critics basically ignored this issue.

39. Crow (1969).

40. Bodmer and Cavalli-Sforza (1970); Dobzhansky (1973b); Lederberg (1973).

41. Hay (1985).

42. Kamin (1974).

43. Cf. Jencks et al. (1973) and Goldberger (1979).

44. Scarr-Salapatek (1976).

45. Lewontin (1970b).

46. Feldman and Lewontin (1975); Kempthorne (1978); Lewontin (1974).

47. As the debate progressed many participants adopted increasingly extreme positions. Jensen, Herrnstein, and Eysenck all started out by claiming to be environmentalist psychologists before advancing increasingly hereditarian positions during the debate. Lewontin had similarly began by claiming that social scientists needed to heed genetics research before arguing, in essence, that human behavior genetics was an impossible science. (See, for example, Jensen [1967; 1972], Lewontin [1970b]; and Harwood [1976]). Many behavior geneticists' narratives of the time talked about becoming increasingly partisan as the controversy ground on (for example, Lykken [2002]; Scarr [1987]). Those who didn't become extreme lost relevance: Dobzhansky who had been a stalwart antiracist geneticist since the 1940s and the thesis advisor of Hirsch, Lewontin, and others—and had helped found behavior genetics on noneugenic grounds—was often ignored in the debate as he tried to forge a moderate position expressing sympathy for Jensen while asserting the value of racial genetic diversity (Dobzhansky [1973b]; Paul [1994]).

48. Friedrichs (1973); Harwood (1977); Lieberman and Reynolds (1978); Sherwood and Nataupsky (1968); Snyderman and Rothman (1988); Tucker (1994).

49. Interviewees #10, #14, #35, and #36. Gottfredson (1994); Lewontin (1976c); Loehlin, Lindzey, and Spuhler (1975); O'Connell (2001); Snyderman and Rothman (1988).

50. Interviewee #34.

51. In our interview Leon Kamin told me that he entered the controversy initially because he wanted to defend Herrnstein's right to speak at Princeton when students threatened to shut down his talk. But then Kamin dismissed this motivation by joking that he was in his "bourgeois liberal" days, suggesting that he now had a more radical sensibility about intellectual freedom.

52. Letter from Richard Lewontin to Sandra Scarr Salapatek, May 8, 1973. Theodosius Dobzhansky Papers, American Philosophical Society. Series I, Box 12, Folder 2.

53. Harwood (1976).

54. Deutsch (1969); Feldman and Lewontin (1975); Goldberger (1979); Hirsch (1975); Jencks (1980); Kamin (1974); Lewontin (1970b; 1974); Morton (1972).

55. Flaschman and Weintraub (1974); Lewontin (1976b); Lewontin, Rose, and Kamin (1984); see also Paul (1987).

56. See, for example, Chase (1980); Gould (1996); Hubbard and Wald (1993); Lewontin, Rose, and Kamin (1984).

57. For example, Trevor Williams responded to a Lewontin (1976a) critique of a "fundamental error" in his article, "most readers will see the reported heritabilities for what they are: modest estimates that may be modestly biased but, in Falconer's words, 'not entirely unrealistic'" (Williams [1977]).

58. For example, Kamin (1977a; 1977b); Lewontin (1976c).

59. For example, Bouchard (1996); Erlenmeyer-Kimling and Jarvik (1963).

60. See Lewontin (1970b; 1976c); also Hirsch (1975).

61. Lewontin (1976c); see also Levins and Lewontin (1985).

62. Although many did critically engage sociobiology when it emerged, seeing in it similar political appeals to biological determinism (Segerstråle [2000]). Economist Arthur Goldberger Goldberger and Kiefer 1989) mentioned that the way data were discussed in the IQ controversy made him more skeptical of how economists report data.

63. Kamin (1974); also see Orbach, Schwartz, and Schwartz (1974) and Schwartz and Schwartz (1974).

64. Jencks et al. (1973, 72) quoted in Paul (1998c, 88).

65. Paul (1998c, 88).

66. Different orientations toward this activity were another reason that Jensen and his allies, but not critics, were incorporated into the field. Both were attempting to redefine behavior genetics, but Jensen and company mostly did so implicitly and were happy to engage behavior geneticists. (While aspects of Jensen's and others' work were treated critically by behavior geneticists, even this critical recognition was a form of legitimation.) Critics, in contrast, explic-

itly tried to define the field, but with indifference to behavior geneticists' evaluations. Perhaps Jensen and his allies didn't integrate fully into behavior genetics, but they eagerly engaged behavior geneticists on their own terms, to submit to behavior geneticists' scrutiny, and to participate positively by doing research that seemed to advance the field's methods and findings. Even if Jensen and the rest were pulling behavior genetics' agenda in directions that some members didn't like, at least they were doing so through their work within the field rather than making pronouncements from beyond its boundaries. If behavior geneticists saw the critics and Jensen's cadre all as invaders, the critics looked like vandals bent on destruction while Jensen, Eysenck, Herrnstein, and their ilk came to look more like industrious immigrants.

67. For example, Cavalli-Sforza and Feldman (1973); Kagan (1969); Kamin (1978); Kempthorne (1978); Layzer (1974); Lewontin (1974); Morton (1972); Schwartz and Schwartz (1974); Vetta (1977)

68. Interviewee #6.

69. Feldman and Lewontin (1975); Goldberger (1979); Jencks (1980).

70. Loehlin, Lindzey, and Spuhler (1975, 99).

71. Feldman and Lewontin (1975); Lewontin (1974).

72. Lewontin (1976c).

73. Lewontin (1974).

74. See Jensen (1980) quoted in Hay (1985, 311). It is worth noting that the cross fostering model still wouldn't eliminate different experiences of racism. Further, the fact that Jensen declared the problem unsolvable hasn't prevented him from continuing to make claims about genetic racial differences (Rushton and Jensen [2005]).

75. Indeed, Jensen's critics would conceive that genes in concert with environments cause individual development and thus differences among individuals and groups. But what they disputed was the notion that we could infer from current conditions that racial IQ differences would persist in some different set of environments (or current environments distributed differently). Block and Dworkin (1974); Hirsch (1978); Lewontin (1974; 1975). Philosopher Ned Block (1996) later pointed out that logically, blacks might be "genetically smarter" than whites in some equalized environments, but current environments reversed that pattern. Block argues that the fact that this possibility was never debated among participants is evidence of the depth of racist assumptions.

76. Dunn and Dobzhansky (1952); Fuller and Thompson (1960); Hirsch (1967c; 1968); Spuhler and Lindzey (1967).

77. Spuhler and Lindzey (1967).

78. This is described in the preface to Loehlin, Lindzey, and Spuhler (1975).

79. Loehlin, Lindzey, and Spuhler (1975, 238).

80. Lewontin (1976c, 95).

81. Lewontin (1976c, 96–97).

82. Quoted in Provine (1986, 879, 880).

83. The Eastern Psychological Association, Society for the Psychological Study of Social Issues, American Anthropological Association, American Linguistics Society, and American Sociological Association all passed similar resolutions (1986, 878).

84. Gottfredson (1994), for example, complains less about attacks than colleagues' cowardice for failing to publicly acknowledge what she believes they know to be true about race.

85. Harwood (1979).

86. Kevles (1985).

87. Goldberger and Kiefer (1989, 144).

88. For example, Lewontin (1970b). Although Lewontin did sometimes insult their competence in genetics and the competence of the editors of *Behavior Genetics* (1976a).

89. See, for example, Goldberger (1976); Gould (1996); Lewontin (1970a; 1970b; 1975).

90. Crow (1969); Morton (1972); Thoday and Parkes (1968).

91. See Paul (1998c, 81–93); and also Francis Crick to John T. Edsall, March 29, 1971. Ernst Mayr to Francis Crick, April 14, 1971. Profiles in Science Collection, National Library of Medicine.

92. Francis Crick to John T. Edsall, March 29, 1971. Profiles in Science Collection, National Library of Medicine. Crick's proposed twins institute incidentally would not have simply searched out separated twins in the world but would actively encourage parents of twins to voluntarily give up one of their twins for adoption and would then trace the results.

93. Ernst Mayr to Francis Crick, April 14, 1971. Profiles in Science Collection, National Library of Medicine. Mayr was deeply concerned that Shockley's apparent racism would fatally stigmatize the eugenic project.

94. Plomin (1979, 420).

95. Lewontin and Cavalli-Sforza (1974, 2).

96. Jacobs et al. (1965); see also Gaylin (1980).

Chapter Four

1. Although various infrahuman species have their own relative advantages, the basic dispute is between those who advocate studying human subjects and those who favor animals.

2. Balaban, Alper, and Kasamon (1996).

3. Benzer (1971).

4. Although scientists have dreamt of affecting adoption placements and encouraging mothers of twins to give one up. See, for example, Vandenberg

(1965, 322) and Francis Crick's proposed "twins institute" mentioned in the last chapter.

5. Merton (1973).

6. Rheinberger (1997). See also Griffiths and Tabery (2008); Tabery (2009).

7. For example, on the left, Lewontin, Rose, and Kamin (1984), and on the right, Snyderman and Rothman (1988).

8. Interviewee #6.

9. Interviewee #2.

10. See Rosenfield (1992) and Klein (1996).

11. Interviewee #7. "Knockout" and "transgenic" studies refer to methods that emerged after the early 1990s upon which this chapter focuses. However, the point about fragmentation and nonrecognition still holds during this period.

12. Interviewees #6, #14, #24, #28, and #31.

13. Abbott (1988); Rose (1992); Scull (1989).

14. Gottesman and Shields (1972; 1967).

15. Cloninger (1986); Cloninger, Rice, and Reich (1979).

16. Plomin (1979).

17. These data were collected in 1979, after most of the fallout from the IQ and race controversy. The differences might have been even greater if people who had already dissociated from the field were included in the roster of older members.

18. BGA (1996).

19. Similar to the University of Minnesota psychology department, which was another leading center of behavior genetics. See table 4.2.

20. Interviewee #35.

21. I develop these ideas further in chapter 5.

22. Interviewee #35.

23. Plomin, DeFries, and McClearn (1980). The current edition is the sixth (Plomin et al. [2013]).

24. Interviewee #7.

25. Interviewee #28. Note that the locution "wouldn't answer on either side of it" indicates the self-perception of being in the middle ground between the field's critics and controversial provocateurs.

26. This isn't an isolated sentiment. One critic (interviewee #9) described an encounter at a meeting with behavior geneticists where after delivering a talk about the dangers of making claims about racial differences in behavior genetics he was accused of being a "coward" for demurring after a behavior geneticist offered to collaborate on a project comparing the IQs of black and white twins.

27. Bourdieu (1975, 23–26).

28. Interviewees #7, #11, and #36.

29. Scarr (1987, 226). For other examples, see Brand (1999); Cattell (1980); Jensen (1972); Lykken (2006); Rushton (1998); Wright (1998).

30. BGA (1985; 1996).

31. Interviewee #35.

32. Interviewee #28.

33. Interviewees #3, #6, #7, #25, #28, and #35.

34. These exchanges included Kamin (1977a; 1977b; 1978); Lewontin (1976a); Munsinger (1977); Williams (1977); Jensen (1976); Schwartz and Schwartz (1976).

35. Interviewee #6.

36. The speaker continued, "people would say 'Oh, we don't want to deal with non-linearity,' and just actively resisted it. They'd just say, that we just don't know enough to do that." Although the decision to ignore these issues was not made cynically (the empirical problem was difficult enough without nonlinearity and other complicating factors), the effect was to produce underqualified and perhaps inflated heritability estimates, to block out other interpretations, and to delay the field's appreciation of these issues for decades.

37. Schiff and Lewontin (1986).

38. As of July 2006 the Institute for Scientific Information's Web of Science had recorded fifty-six citations to Schiff and Lewontin (1986). Of these, only eighteen were made by people who do behavior genetics. However, eleven citations are in British psychiatrist Michael Rutter's work, four were psychologist and behavior genetics dissident Eric Turkheimer's work. Only three of the other behavior genetics references to this work did not directly involve one of these two authors (and at least one of these three involved Rutter's coauthors).

39. Hirsch (1975; 1981); Hirsch, Beeman, and Tully (1980a); Hirsch, McGuire, and Vetta (1980b); Hirsch and Tully (1982).

40. Interviewee #34.

41. Interviewees #14, #34, and #35. After Hirsch's death in 2008, obituaries tried to reaffirm his centrality to behavior genetics (McGuire [2008]; Roubertoux [2008]; Wahlsten [2008]).

42. Interviewee #7 (part 2).

43. Interviewee #7 (part 2).

44. Interviewee #7 (part 1).

45. Interviewee #7 (part 1).

46. Similar complaints from animal researchers about the editorial and peer review processes stymying critiques of behavior genetics claims (especially those made by Arthur Jensen) have been published (see, for example, Kempthorne [1978]; Schönemann [2001]). These have suggested that the disinclination to publish critiques is a broad problem in scientific journals that is not limited to those focusing on behavior genetics.

47. Arvey and al. (1994).

48. Interviewee #35.

49. Despite the image of science as animated by unbridled "conjecture and

refutation" (Popper [1962]), sociologists have shown that science would be dysfunctional without tacit and explicit norms curtailing and limiting mutual criticism (Collins [1985]; Shapin [1994]).

50. Interviewee #7 (part 2).

51. Critical forums on behavior genetics have recently been hosted in external journals like *Cahiers de Psychologie Cognitive* (1990), *Genetica* (1997), and *Human Development* (1995, 2006).

52. Interviewee #7 (part 2).

53. Interviewee #11.

54. Merton (1973); Popper (1962).

55. Bourdieu (2004); Collins (1985); Shapin (1994).

56. Interviewee #28.

57. Interviewee #6.

58. Interviewee #7.

59. Interviewee #3.

60. Interviewee #3.

61. Interviewee #11.

62. Interviewee #35.

63. Gadamer (2004).

64. Interviewee #34.

65. Bourdieu (2004).

66. BGA (1996). Eleven did not record what species they study.

67. See http://bga.org/pages/52/Historical_table_of_BGA_Meetings.html, accessed June 1, 2012. Four of the presidents and Dobzhansky winners have done both animal and human research.

68. As of August 17, 2006, these included C. Robert Cloninger, Lynn Eleanor DeLisi, Lindon J. Eaves, Steve V. Faraone, Elliot S. Gershon, Andrew Charles Heath, Kenneth Kendler, Kenneth K. Kidd, Sarnoff A. Mednick, Kathleen R. Merikangas, Terrie E. Moffitt, Michael C. Neale, Robert Plomin, Michael L. Rutter, and Ming T. Tsuang. ISI's list of Highly Cited researchers is constantly updated.

69. These include the neuroscientists Eric Kandel and Donald Pfaff and the geneticist Seymore Benzer. For example, I had been told by one of my interviewees that Eric Kandel, a Nobel laureate focusing on the neurogenetics of memory in flies, was always willing to be interviewed. He responded to my requests with brief but firm rebuffs: "This is not my area of expertise. Sorry." Although are any number of reasons to be wary of being interviewed by a sociologist, his rebuff is also a sign of low (or no) commitment to behavior genetics—a sense that it is just not worth getting involved.

70. Interviewee #28. I did a search of behavior genetics articles that found only about 15 percent mentioned race, gender, ethnicity, or class. Most were mentioned only to specify the study population; only about one-third, or 5 percent,

of the total said anything about group differences. Of these, most were modest gender differences or ethnic variation in alcohol sensitivity. The audit used the online databases Biological Abstracts, Medline, and Ovid for research done in 2001 and earlier. A keyword search was done for the terms *gene* and *homosexual, violence, violent, intelligence, mental illness, mental disorder, substance abuse,* or *aggression* (alcoholism was excluded because it swept up too many irrelevant studies). The searched picked up 583 articles, of which I deemed 186 irrelevant to behavior genetics after reading the titles and abstracts. Of those that remained, seventy-four mentioned race, ethnicity, gender, or class, and of those about one-third (~25) mentioned group differences. The uncertainty comes from difficulty I had interpreting in some abstracts whether the point of mentioning group identities was to make group comparisons or simply to label a population.

71. For example, Dobzhansky (1962); Freedman (1968); Fuller (1983); Ginsburg and Laughlin (1966).

72. Bouchard et al. (1990); Segal (1993).

73. Scarr (1987, 228).

74. Such justifications were risky in general because they gave critics like Feldman and Lewontin an additional point of entry (Feldman and Lewontin [1975]). They argued that an evolutionarily relevant trait will experience selection that limits variability in the population. If there is substantial variation in the traits that behavioral geneticists care about, like IQ, then they argued that trait could not have been crucially relevant over evolutionary time. If high IQ is selectively advantaged, what explains the persistence of low IQ?

Prima facie then, the claims of evolutionary importance and genetic importance (i.e., high heritability) seemed to pull in opposite directions. Even if there was a case to be made for their compatibility, doing so risked pulling behavior geneticists with their mainly psychological expertise further into the territory of evolutionary biology.

75. Dawkins (1976); Wilson (1975).

76. Gould (1996); Lewontin, Rose, and Kamin (1984); see also Segerstråle (2000).

77. McClearn and DeFries (1973).

78. Layzer (1974); Lewontin (1974).

79. Jensen (1973). See also Tabery (2009).

80. Plomin, DeFries, and Loehlin (1977). DeFries, interestingly, was mainly an animal behavior geneticist at the time.

81. Interviewee #7. On the history of ethology, see Burkhardt (1981).

82. Interviewee #14.

83. Freedman (1965; 1968).

84. Ginsburg and Laughlin (1966).

85. Royce and Mos (1979).

86. Interviewee #7 (part 2). Later in the interview, he continued by referring

to a target article by animal behavior geneticist Douglas Wahlsten (1990) in the debate-oriented journal *Behavioral and Brain Sciences* about the limitations of detecting gene/environment interactions in heritability estimation techniques: "And even in *Behavioral and Brain Sciences*, I didn't have the feeling at that point that there was any communication going on any more. It's completely ritualized, restating of old positions."

87. Interviewee #35.

88. Horn, Loehlin, and Willerman (1979, 198).

89. Kaplan (2000, 37). Other such anomalies have been described by Kamin and Goldberger (2002) and Joseph (2004; 2006; 2011).

90. Interviewee #7 (part 1).

91. Scott (1985, 417).

92. Crabbe, Wahlsten, and Dudek (1999). See also Enserink (1999).

93. Indeed, the finding seems to argue against the environmental determination of behavior either, at least against the determination by known or knowable causes. It suggests a strong role of stochasticism.

94. Griffiths and Tabery (2008).

Chapter Five

1. In principle, nature versus nurture and reductionism vs. emergence are really separate dimensions. For example, a nurture explanation could also emphasize reductionistic mechanisms. Arguably this is what research in epigenetics is beginning to show—the nurturant environment can cause molecular changes affecting the expression of DNA that can alter behavior and be inherited intergenerationally. Despite the ways that epigenetics and related research seem to be muddling the relationship, historically the two dyads, nature and reductionism and nurture and emergence, have been closely associated.

2. Abbott (1988); Collins (2000); Frickel and Gross (2005); Gieryn (1999).

3. For a sample of recent critiques, see Chabris et al. (forthcoming); Charney and English (2012); Horwitz et al. (2003); Joseph (2011); Kaplan (2006); Lerner (2006).

4. Collins (1985); Gieryn (1999); Rudwick (1985); Shapin and Schaffer (1985); see also Adut (2008). Collins (2000) and Simon (2002) discuss examples where discredited sciences are able to persist, but this is largely by avoiding engagement, especially conflict, with the scientific mainstream—the opposite of what behavior genetics did.

5. Scarr (1987, 224–26).

6. Aldhous (1992, 164).

7. Interviewee #25.

8. Interviewee #35. The speaker also noted that it was almost always "boys,"

and that women tended not to enter the field. Sandra Scarr, however, is a woman who embodied this swagger as much as anyone.

9. Interviewee #28.

10. Scarr (1992).

11. Scarr (1992, 3, emphasis in the original). Here Scarr engages in a subtle slip akin to those she would accuse in her critics. "Individual differences" refers to a population trait, but "development" refers to an individual process. Individual traits develop, and they turn out to be differences only in comparison to others.

12. Scarr (1992, 10–14).

13. Baumrind (1993); Jackson (1993).

14. Scarr (1993).

15. Scarr (1993, 1342).

16. Scarr (1993, 1343).

17. Scarr (1993, 1345).

18. Scarr (1993, 1346).

19. In the current era of epigenetics, Lamarck has moved from being a punch line to the scientific vanguard.

20. Kamin (1974; 1981); Lewontin, Rose, and Kamin (1984). See also Joseph (1998; 2006).

21. Scarr and Carter-Saltzman (1979). Behavior geneticists continued to use this technique; see Kendler et al. (1993).

22. Bouchard et al. (1990).

23. Bouchard et al. (1990); Lykken et al. (1990).

24. Beckwith and Alper (1998); Joseph (2006).

25. Baumrind (1993); Jackson (1993); Stoolmiller (1999); Taylor (1980).

26. Plomin and DeFries (1985); Rhea and Corley (1994), Scarr (1992).

27. Scarr (1992).

28. Alper and Beckwith (1994); Seligman (1994).

29. Thomas and Thomas (1928, 572).

30. Kamin (1974).

31. On the Colorado project, see http://ibgwww.colorado.edu/cap/history .html, accessed April 24, 2012. On Texas, see Horn (1983). On Minnesota, see Holden (1980) and Wright (1998). On SATSA, see http://ki.se/ki/jsp/polopoly .jsp?d=13903&a=30148&1=en, accessed July 16, 2012.

32. Wright (1998, 39–41).

33. Loehlin and Nichols (1976).

34. Dusek (1987).

35. This static, dispositional conception of behavior had similarities to the way eugenicists and early twentieth-century psychologists conceived of "character."

36. Scarr (1981, 528–29).

37. Indeed, in many of the basic designs of twin, adoption, and family stud-

ies even these basic contextual variables are not measured per se. Rather, by comparing the correlations between different types of kin relationships in the same or different households, these contextual factors are "controlled" in various ways, and their effects can be assigned in the statistical models to environmental factors family members experience as "shared" (e.g., parents' education or income) and "nonshared" (e.g., activities and experiences outside the home). One factor in this preference may have been the difficulty and expense of collecting data beyond responses to paper-and-pencil tests from widely dispersed twins and adoptees.

38. Gottlieb (1995); Jackson (1993); Taylor (1980); Wachs (1983).

39. See, for example, Anastasi (1958) and Freedman (1965). Scott (1985) listed motivation, opportunities for escape, and adaptation in addition to genetic heterogeneity in his model of animal behavior.

40. In structural equation modeling the analyst hypothesizes a set of causal relationships between social factors. These are expressed as a set of interconnected equations (in some computer programs drawn as a kind of flow chart). Then empirical data (often correlations among hypothesized factors) are compared to the hypothetical scheme using statistical techniques to see how well they fit it.

41. Neale and Cardon (1992).

42. Plomin et al. (2008, 307–8).

43. Plomin and Daniels (1987); Rowe (1994); Scarr (1992).

44. McGue et al. (1993).

45. As interviewee #35 put it: "Things start to get cut up more and more finely with more and more qualifications and interactions and, 'Well, it seems like it works this way in women in Finland, but in this set of Irish men it seems like it works that way,' . . . you never seem to reach any scientific rock bottom."

46. Detterman, Thompson, and Plomin (1990).

47. Interviewee #7 (part 1).

48. Interviewee #7 (part 1).

49. Lakatos (1970).

50. See Kuhn (1970), especially chapter 4.

51. Turkheimer (2000). Although psychiatric geneticist Michael Rutter (2002) disputes the "law" about the noneffect of shared environment.

52. Urbach (1974a; 1974b) argued the hereditarian argument in the IQ debate was progressive and their opponents' was degenerating. But Tizard (1976) pointed out that this was only true if one accepted not only the hereditarian arguments but also their assumptions and framing of the problem.

53. Interviewee #35.

54. Quoted in Kamin and Goldberger (2002, 93).

55. Interviewee #28, emphasis added.

56. When I attended the Behavior Genetics Association annual meeting in

2002, I struck up a conversation with a vendor of analytical software tools. I mentioned that I had heard that some geneticists do not consider behavior geneticists to be expert geneticists. His perception was that their genetic analysis is quite sophisticated, but as a social psychologist he thought their measures of behavior, in particular the reliance on paper-and-pencil psychometric tests, was relatively crude by psychology's standards.

57. Rose (1992).

58. Abbott (1988).

59. Bourdieu (1975; 2004).

60. Psychology tests are used as tools for sorting and selecting in schools, businesses, military, and other institutions. Culturally, psychological concepts have become much of the vernacular by which we conceive of ourselves and our relationships.

61. Interviewee #6.

62. Interviewee #28.

63. Panofsky (2011).

64. For the many biology- and development-oriented fields of animal behavior, behavior genetics has offered few tools as well as a long history of politically uncomfortable controversy. These fields have tended to ignore behavior genetics, which is why those animal researchers who have committed to behavior genetics feel scientifically isolated.

65. Interviewee #28.

66. See Kagan (1969) and Wachs (1983) and later Kagan (2003) and Wachs and Plomin (1991).

67. Plomin et al. (2003a, 10).

68. Interviewee #35.

69. Interviewee #28. Interviewees #14, #25, #31, and #35 told similar stories. Such conversion stories appear to involve a dramatic shift of position, but in many cases they may actually represent a consistent provocative disposition but a switch in the substantive means for provocation.

70. Condit (1999); Conrad (1997); Nelkin and Lindee (2000).

71. Nelkin (1987).

72. Interviewee #28.

73. Interviewee #28.

74. Rowe (1994).

75. Rowe (1994, 224). He is citing from Lewontin, Rose, and Kamin (1984, 282), but could have been talking about many psychologists who have made the same charge. See, for example, Gottlieb (1995) or Kagan (1984).

76. Rowe (1994, 222).

77. Rowe (1994, 223).

78. For example, Loehlin, Lindzey, and Spuhler (1975).

79. Plomin and DeFries (1985); Scarr (1992).

80. Stoolmiller (1999). This article was published slightly later than the period of focus in this chapter, however it was submitted first in 1997 and it draws on materials available contemporary to the early 1990s.

81. Scarr (1992; 1993).

82. Dickens and Flynn (2001).

83. For other examples, see R. Rose (1995).

84. Interviewee #35.

85. Adut (2008).

86. Hughes (1991).

87. Interviewee #35. See also Panofsky (forthcoming).

Chapter Six

1. Brown and Michael (2003).

2. Some quantitative genetic models allow researchers to make inferences about the mode of transmission (e.g., additive, dominance, or interactive effects). But none of these specify genomic locations or molecular functions.

3. Lewontin (1974).

4. Joseph (2011).

5. Alper and Beckwith (1994); Billings, Beckwith, and Alper (1992); Duster (1990; 2004), Nelkin and Lindee (2000); Rose (1998).

6. See Beckwith (1991, 1–2).

7. Gilbert (1992, 86).

8. Jaroff (1989).

9. Watson (2003, xxii).

10. Koshland (1987; 1989; 1990).

11. Hamer and Copeland (1994, 25). Chapter 7 considers behavior genetics and the pursuit of fame and notoriety.

12. National Institutes of Mental Health (2006).

13. The story is told more fully in Panofsky (2011).

14. Shorter (1997).

15. Interviewee #24.

16. Kendler (1994, 100).

17. I obtained the data for figure 6.1 using a keyword search for *behavioral genetics* (the NIH's database label for the field) on new (nonrenewal) research grants (eliminating other forms of funding support) in the NIH's "CRISP Database" (http://crisp.cit.nih.gov/). Funding data are difficult to obtain for scientific subfields (especially when keywords associated with the subfield are not highly specific).

18. NIH has historically not provided funding data with the CRISP database. (This is changing with the new RePORT system.) Data for figure 6.2 were ob-

tained from an online resource from an advocacy group called The Sunshine Project. They combined NIH's budgetary disbursement database with its grant database between 1996 and 2005 in an online resource called "CRISPer," the project, now defunct, was available at www.cbwtransparency.org/crisper.

19. The organization was originally called the European Behavioural and Neural Genetics Society, but to broaden its reach was renamed International at the first conference in 1997. See http://www.ibangs.org/index.php?option=com _content&view=article&id=33, accessed September 13, 2012.

20. Membership rosters were published in 1985 and 1996; aggregate numbers are published in the BGA's minutes in *Behavior Genetics*.

21. This is a ballpark estimate. Circa 2011–12, BGA membership was down to 444, ISPG was 550, and IBANGS was 228.

22. Egeland et al. (1987).

23. Kelsoe et al. (1989).

24. Robertson (1989).

25. Risch and Botstein (1996, 351).

26. Kato (2007).

27. Hamer et al. (1993).

28. D. Miller (1995).

29. Hamer and Copeland (1994).

30. Rice et al. (1999). See also Mustanski et al. (2005).

31. Blum et al. (1990).

32. Holden (1994).

33. Conrad and Weinberg (1996).

34. Caspi et al. (2003).

35. News Editorial Staffs (2003).

36. Risch et al. (2009).

37. Zammit, Wiles, and Lewis (2010a); Zammit, Owen, and Lewis (2010b).

38. Bouchard (2009); Deary, Johnson, and Houlihan (2009).

39. Davies et al. (2011).

40. Personal communication, September 25, 2011. The Associated Press covered the story; see Chang and Ritter (2011).

41. Hirschhorn et al. (2002).

42. Bosker et al. (2011); Buxbaum, Baron-Cohen, and Devlin (2010); Charney and English (2012); Newton-Cheh and Hirschhorn (2005); Siontis, Patsopoulos, and Ioannidis (2010); Sullivan (2007).

43. Gorroochurn et al. (2007); van den Oord and Sullivan (2003).

44. For example, Kaplan (2000); Moore (2001); S. Rose (1995).

45. Marshall (1994, 1694).

46. Joseph (2006).

47. Kevles (1985).

48. Rutter and Plomin (1997, 214).

49. Plomin et al. (2003a, 9).

50. Stoltenberg and Burmeister (2000, 927) call this equifinality—where different causal paths can produce the same outcome.

51. Plomin, Owen, and McGuffin (1994).

52. Gottesman and Gould (2003).

53. Interviewee #35.

54. Rogaev (2012).

55. Sullivan (2011).

56. See Arribas-Ayllon, Bartlett, and Featherstone (2010).

57. Arribas-Ayllon, Bartlett, and Featherstone (2010, 516–17).

58. Arribas-Ayllon, Bartlett, and Featherstone (2010, 517).

59. Eric Turkheimer (2011; 2006) stands out for his meditations on failure while Kenneth Kendler (2005) offers a slightly more optimistic vision.

60. Interviewee #35.

61. Bourdieu (2004, 62–71).

62. McGue (2008); Plomin et al. (2003b).

63. Arribas-Ayllon, Bartlett, and Featherstone (2010, 515–16).

64. Bourdieu (2004) and Fligstein and McAdam (2012) emphasize conflicts between incumbents and challengers in their theories of field change and stasis. Incumbents often control resources and are able to fend off challenges to reproduce a field structure that benefits them. Behavior genetics is an interesting twist in that molecular genetics challengers typically had greater resources, yet their actions helped to reproduce the structure of the field and incumbent behavior geneticists' position within it because the challengers declined to compete for the incumbents' positions and authority over the field.

65. Interviewee #13.

66. Interviewee #28.

67. These are molecular genetics techniques where a particular gene is either deleted (knocked out) or added to the genome of an animal embryo, that animal is allowed to develop, and the impacts of the genetic change are observed.

68. Interviewee #7 (part 1).

69. That animal behavior geneticists should particularly resent such incursions is not surprising. They, after all, were the ones who tended to pine for a well-functioning field and to lament their lack of control of its norms. Furthermore, they experienced these incursions as a loss of control that they formerly had. One interviewee (#34) claimed that Jerry Hirsch had essentially driven Seymore Benzer, an eminent researcher of genetic mutations in fruit flies, out of the field. In 1974, Benzer published a claim that he had demonstrated associative learning in fruit flies and a genetic mutant, *dunce*, that couldn't learn (Quinn, Harris, and Benzer [1974]). Hirsch, whose lab had tried and failed to establish learning in flies for twenty years, didn't believe the claim, and he wrote a scathing critique that he distributed to Benzer's colleagues. According to the inter-

viewee, Benzer could not adequately defend his position and turned away from behavioral research in flies. We might note three things here: First, molecular geneticists' hubris predates the current era. Second, Benzer and other molecular geneticists get credit for discoveries that behavior geneticists feel they deserve. And third, in an earlier era when the field was not yet so fragmented and animal researchers were more powerful, behavior geneticists had more authority to police the field's boundaries.

70. Plomin (1990, 188). Joseph (2011) has tracked Plomin's public enthusiasm for molecular genetics and his struggles to come to grips with its failures.

71. Plomin et al. (2003b).

72. Interviewee #7 (part 2).

73. Charney and English (2012). These figures are from a detailed online supplement accompanying the article: http://tinyurl.com/9ry9542, accessed September 12, 2012.

74. Interviewee #3. Since this interview, Dean Hamer has published a GWAS study of homosexuality. See Mustanski et al. (2005).

75. This is equally true for animal and human research. The proportion of studies that mention at least one popular model organism and one of the molecular keywords is about the same as the overall proportions.

76. Eighth Annual International Summer School on Behavioral Neurogenetics, Worcester, Massachusetts, August 5–9, 2002.

77. Arribas-Ayllon, Bartlett, and Featherstone (2010).

78. Kaplan (2000).

79. Maher (2008).

80. Haworth and Plomin (2010).

81. Interviewee #28.

82. Miele (2008, 37).

83. Turkheimer (2004; 2006).

84. Interviewees #24, #25, and #30.

85. Interviewee #25.

86. Interviewee #25.

87. Turkheimer et al. (2003).

88. Interviewee #25.

89. Dick et al. (2001); Dick and Rose (2002).

90. Caspi et al. (2002).

91. Caspi et al. (2003, 389).

92. Gieryn (1983, 787).

93. This suggests that in addition to social changes opening up new territories in an intellectual space of possibility, they might also work by unlocking fractal patterns of differentiation within narrower and narrower spaces (Abbott [2001]).

94. Helmreich (2003).
95. Interviewee #35.

Chapter Seven

1. Scarr (1987, 227).
2. Nelkin and Lindee (2000).
3. Lippman (1992); Nelkin and Lindee (2000).
4. Barnes and Dupré (2008, 142).
5. Interviewee #15. This comment was made specifically in reference to Carson and Rothstein (1999).
6. Page (1972).
7. Hirsch (1981, 10). Historian Robert Tucker (1994, 276) showed that when Jensen later reviewed Dobzhansky's *Genetic Diversity and Human Equality*, he expressed irritation that the book would even bring up the idea of racist misuses of science.
8. Scarr (1987, 221).
9. Johnson and McGue (2010).
10. Lykken (2006, 34–40).
11. Moore (2008).
12. Arvey et al. (1994); Bouchard (1995).
13. Interviewees #14 and #35. See also Gottfredson (1997). Gottfredson (1994) charges behavior geneticists and psychologists with cowardice for not lending race researchers more support.
14. Hirsch (1997b). Hirsch (1997b) introduced a special issue of *Genetica* featuring critiques of *The Bell Curve*.
15. Interviewee #23.
16. Interviewees #2, #7 (parts 1 and 2), #14, #23, #29, #35, and #36. Unfortunately, I do not have access to this archive.
17. Heath (1995, 590).
18. Hunt (2010, 264); Loehlin (1992).
19. Beckwith (2001).
20. Gottfredson (2010).
21. Interviewees #29 and #35, for example.
22. Holden and Bouchard (2009); Scarr (1987); Wright (1998).
23. Sherman et al. (1997, 1273).
24. Interviewee #13.
25. Interviewee #3.
26. Interviewee #35.
27. Breggin (1995).

28. Masters (1996).

29. Parens (2004, S5).

30. Beckwith (2001, 193).

31. Interviewee #5. This was stated specifically in reference to Nelkin and Lindee's *DNA Mystique* (1995), but it is appropriate here.

32. This is my interpretation after having observed two of the five AAAS/Hastings meetings (November 14–16, 2001, and June 10–12, 2002). See also Parens (2004); Parens, Chapman, and Press (2006).

33. See, for example, the description of Maryland conference organizer David Wasserman (1996).

34. Nuffield Council on Bioethics (2002).

35. Hay (2003).

36. Wahlsten (2003).

37. Interviewees #2, #9, #26, #35.

38. Rabinow and Bennett (2012).

39. Hamer and Copeland (1994, 27).

40. Evans et al. (2005); Mekel-Bobrov et al. (2005).

41. Balter (2006); Rice et al. (1999).

42. Richardson (2011).

43. Harris (1998); Pinker (2002); Ridley (2003); Wright (1998). See also Burr (1996); Shenk (2010); Weiner (1999).

44. Gould (1996); Hubbard and Wald (1993); Lewontin, Rose, and Kamin (1984). See also Moore (2001); Nisbett (2010).

45. Cassidy (2005; 2006).

46. Herrnstein (1973); Herrnstein and Murray (1994); Rushton (1994); Wilson and Herrnstein (1985). See also Hamer (2004).

47. Rowe (1994); Rutter (2006). See also Clark and Grunstein (2000).

48. Interviewee #11.

49. Interviewee #14.

50. See chapter 3 and Provine (1986); Tucker (1994).

51. See Provine (1986) and Wright (1986); also Kevles (1985).

52. Fischer et al. (1996); Fraser (1995); Jacoby and Glauberman (1995); Kincheloe, Steinberg, and Gresson (1996); Neisser et al. (1996).

53. Devlin et al. (1997).

54. Botstein (1997, 212).

55. Beckwith (2002, 201–6).

56. Allen et al. (1996).

57. Beckwith (2002, 200).

58. Burr (1996, 274). Incidentally, Lewontin has never been quiet about his political opposition to behavior genetics (as well as his technical challenges). Botstein seems to miss Lewontin's larger claim that it isn't just critique that is politically motivated—so is behavior genetics and any other science. For Lewontin,

the question of why a scientific question is or isn't worth asking is always a political question. Later in the interview Botstein seems to agree with Lewontin that the political reasons for seeking out the "gay gene" are poor.

59. Hirschman (1970).

60. Aldhous (1992, 165).

61. Weber (1946, 135).

62. Weber (1946, 137). Weber, of course, disparaged these aspirations.

63. Hamer and Copeland (1994, 19).

64. See Hamer (2004); Hamer and Copeland (1998); and http://wpsu.org/outinthesilence/, accessed August 16, 2012.

65. Quoted in Richardson (2011, 441).

66. Balter (2006); Richardson (2011).

67. Larson (2013); Naik (2013); Yong (2013).

68. Larson (2013, 37); Yong (2013, 299).

69. Interviewee #3.

70. Sherman et al. (1997, 1274).

71. Condit (1999); Conrad (2001); D. Miller (1995); Nelkin and Lindee (2000); Snyderman and Rothman (1988).

72. Bucchi (1996).

73. Holden (1980, 1324).

74. Dusek (1987). See also Segal (2012).

75. Seigel (1999). See also Ebstein et al. (1996).

76. The claim's timeliness and boldness got it global play far beyond the *Jerusalem Post*, where it first appeared.

77. Journalists' desire to present provocative ideas is another important dynamic, though beyond my purview here. Cassidy (2005) has noted that frequently evolutionary psychology is presented by reporters not on the science beat. Speaking directly about human nature seems to appeal to journalists too.

78. Interviewee #7 (part 2).

79. http://www.pioneerfund.org/Board.html, accessed August 20, 2012.

80. D. Miller (1995).

81. Rowe (1997).

82. Dusek (1987). Another charge is that the notoriety and publicity has attracted twins motivated to emphasize their similarities, thus biasing heritability estimates upward (see Segal [2012, 306–7]).

83. Segal (2012, 309–19, appendix A).

84. I described funding patterns in chapter 1.

85. Interviewee #23.

86. Sternberg (1985).

87. Blue (2007).

88. Interviews #3, #11, #25, #34. Also Rutter (2002).

89. Elias and Dunning (1994).

90. Lippman (1991).

91. Barnes and Dupré (2008).

92. Nelkin and Lindee (2000). See also Lippman (1992).

93. Starr (1982).

94. Rose (1992).

95. Kirp (2007).

96. Brown and Josephs (1999); Steele and Aronson (1995).

97. Baumrind (1993, 1312).

98. Jackson (1993, 1328). See also Kaplan (2000, 170–85).

99. Even the "gloomy prospect" argument in behavior genetics, which argues against genetic determinism, but also claims that the causes of behavior may be so complex as to be effectively unknowable, might promote dangerous pessimism. See Turkheimer (2006).

100. Nelkin (1999).

101. Harwood (1979).

102. D. Miller (1995). Some legal arguments about whether to overturn the California gay marriage ban, Prop. 8, turned on the discussion of whether homosexuality is genetic, therefore immutable and subject to legal protection; see Farrell (2010).

103. Aspinwall, Brown, and Tabery (2012).

104. The vignette explicitly said rehabilitation was not an option. The authors rightly suggest that research reflecting a range of biological accounts beyond hard determinism be similarly tested.

105. Shostak et al. (2009).

106. Clearly, more research is warranted in this area, both about general opinion and experts (such as teachers and therapists). In particular, do geneticized understandings lead to fatalism or to preferences for increased investment in training or rehabilitation?

107. Fukuyama (2002); Sandel (2007).

108. Caspi et al. (2002); Caspi et al. (2003).

109. See Panofsky (2009) for a fuller critical discussion of deploying behavior genetics in social policy.

110. Hirschman (1991).

111. Harvey (2005, 2).

112. Rowe (1994, 188–89).

113. Interviewee #34.

114. Scarr (1987, 227).

115. Rowe (1994, 223, emphasis added).

116. Interviewees #23 and #35.

117. Hedgecoe (1998).

118. Bucchi (1998); Greenberg (2001).

119. Bliss (2012).

120. Rabinow and Bennett (2012).

121. Gibbons et al. (1994).

Conclusion

1. Camic (2006); Mialet (2003); Sismondo (2011).

2. For a similar approach, see Medvetz (2012); Stampnitzky (2013).

3. Alexander (1995); Lamont (1992).

4. Knorr-Cetina (1999); Lamont (2009); Moore (2008); Popp Berman (2012); Rheinberger (1997).

5. See, for example, Kaplan (2000).

6. Sennett (1980). See also Panofsky (forthcoming).

7. Bliss (2012); Epstein (2007); Moore (2008); Shapin (2008).

8. Bourdieu (1990); Taylor (1993).

9. Frickel et al. (2010); Proctor and Schiebinger (2008).

10. Shapin (2008).

11. Bliss (2012); Frickel (2006).

12. Couldry (2003).

13. Popp Berman (2012).

14. Krimsky (2003); Mirowski (2011).

15. Kirp (2003).

16. Oettl (2012).

17. Mirowski and Van Horn (2005); Sismondo (2009).

18. Gibbons et al. (1994); Jacobs and Frickel (2009).

19. INSPIRE is Integrated NSF Support Promoting Interdisciplinary Research and Education. IGERT is Integrative Graduate Education and Research Traineeship. CREATIV is Creative Research Awards for Transformative Interdisciplinary Ventures. See www.nsf.gov/od/oia/additional_resources/interdisciplinary_research/index.jsp, accessed April 30, 2013.

20. Jacobs and Frickel (2009, 46).

21. Turner (2000).

22. Moore (2008).

23. Bosk (2008); Evans (2012).

24. I am simplifying the complicated story told earlier and do not want to imply that behavior geneticists acted unilaterally or "unscientifically" in any simple sense.

25. Evans (2002).

26. Bliss (2012).

27. Richardson (2011).

28. Interviewees #3, #7, #35, #36.

29. "The project of social theory that undertakes simultaneously critique

of received categories, critique of theoretical practice, and critical substantive analysis of social life in terms of the possible, not just the actual" (Calhoun [1993, 63]).

30. Kaplan (2000); Lewontin, Rose, and Kamin (1984); Nelkin and Lindee (1995).

31. Frickel et al. (2010).

32. Ridley (2003); Rutter (2006). Even "interactionist" models like Caspi and Moffitt's conceive of nature and nurture as separable even if they portray them as covarying.

33. Griffiths and Tabery (2008); Kaplan (2000); Tabery (2009).

34. Barnes and Dupré (2008); Charney (2012); Keller (2010).

References

Abbott, Andrew Delano. 1988. *The System of Professions: An Essay on the Division of Expert Labor.* Chicago: University of Chicago Press.

———. 2001. *Chaos of Disciplines.* Chicago: University of Chicago Press.

Ad Hoc Committee on Genetic Factors in Human Performance. 1972. "Recommendations with Respect to the Behavioral and Social Aspects of Human Genetics." *Proceedings of the National Academy of the Sciences* 69: 1–3.

Adut, Ari. 2004. "Scandal as Norm Entrepreneurship Strategy: Corruption and the French Investigating Magistrates." *Theory and Society* 33: 529–78.

———. 2008. *On Scandal.* New York: Cambridge University Press.

Albert, Mathieu, Suzanne Laberge, and Brian D. Hodges. 2009. "Boundary-Work in the Health Research Field." *Minerva* 47: 171–94.

Aldhous, Peter. 1992. "The Promise and Pitfalls of Molecular Genetics." *Science* 257: 164–65.

Alexander, Jeffrey. 1995. *Fin de Siècle Social Theory: Relativism, Reduction, and the Problem of Reason.* London: Verso.

Allen, A., B. Anderson, L. Andrews, J. Beckwith, J. Bowman, R. Cook-Deegan, D. Cox, T. Duster, R. Eisenberg, B. Fine, N. Holtzman, P. King, P. Kitcher, J. McInerney, V. McKusick, J. Mulvihill, J. Murray, R. Murray, T. Murray, D. Nelkin, R. Rapp, M. Saxton, and N. Wexler. 1996. "The Bell Curve: Statement by the Nih-Doe Joint Working Group on the Ethical, Legal, and Social Implications of Human Genome Research." *American Journal of Human Genetics* 59: 487–88.

Allen, Garland E. 1996. "The Double-Edged Sword of Genetic Determination: Social and Political Agendas in Genetic Studies of Homosexuality, 1940–1994." In *Science and Homosexualities,* edited by V. A. Rosario, 242–70. New York: Routledge.

———. 1998. "Genetics and Behavior." In *Encyclopedia of Applied Ethics,* vol. 2: 435–43. San Diego: Academic Press.

———. 1999. "Modern Biological Determinism: The Violence Initiative, the Human Genome Project, and the New Eugenics." In *The Practices of Human Genetics*, edited by M. Fortun and E. Mendelsohn, 1–23. Dordrecht: Kluwer Academic Publishers.

Alper, Joseph S., and Jonathan Beckwith. 1994. "Genetic Fatalism and Social Policy: The Implications of Behavior Genetics Research." *Yale Journal of Biology and Medicine* 66: 511–24.

Anastasi, Anne. 1958. "Heredity, Environment, and the Question 'How?'" *Psychological Review* 65: 197–208.

Anderson, Benedict. 1983. *Imagined Communities: Reflections on the Origin and Spread of Nationalism*. London: Verso.

Arribas-Ayllon, Michael, Andrew Bartlett, and Katie Featherstone. 2010. "Complexity and Accountability: The Witches' Brew of Psychiatric Genetics." *Social Studies of Science* 40: 499–524.

Arvey, Richard D., et al. 1994. "Mainstream Science on Intelligence." *Wall Street Journal*, A19.

Arvey, Richard D., B. P. McCall, T. J. Bouchard, Paul Taubman, and M. A. Cavanaugh. 1994. "Genetic Influences on Job Satisfaction and Work Value." *Personality and Individual Differences* 17: 21–33.

Aspinwall, Lisa G., Teneille R. Brown, and James Tabery. 2012. "The Double-Edged Sword: Does Biomechanism Increase or Decrease Judges' Sentencing of Psychopaths?" *Science* 337: 846–49.

Bajema, Carl. 1968. "Relation of Fertility to Occupational Status, IQ Educational Attainments, and Size of Family Origin: A Follow-Up Study of Male Kalamazoo Public School Population." *Eugenics Quarterly* 15: 198–200.

Baker, Catherine. 2004. *Behavioral Genetics: An Introduction to How Genes and Environments Interact through Development to Shape Differences in Mood, Personality, and Intelligence*. Washington, DC: AAAS.

Balaban, Evan, Joseph S. Alper, and Yvette L. Kasamon. 1996. "Mean Genes and the Biology of Aggression: A Critical Review of Recent Animal and Human Research." *Journal of Neurogenetics* 11: 1–43.

Balter, Michael. 2006. "Links between Brain Genes, Evolution, and Cognition Challenged." *Science* 314: 1872.

Barkan, Elazar. 1993. *The Retreat of Scientific Racism: Changing Concepts of Race in Britain and the United States between the World Wars*. Cambridge: Cambridge University Press.

Barnes, Barry, and John Dupré. 2008. *Genomes and What to Make of Them*. Chicago: University of Chicago Press.

Baughman, Robert W., Rebecca Farkas, Marlene Guzman, and Michael F. Huerta. 2006. "The National Institutes of Health Blueprint for Neuroscience Research." *Journal of Neuroscience* 26: 10329–31.

Baumrind, D. 1993. "The Average Expectable Environment Is Not Good Enough: A Response to Scarr." *Child Development* 64: 1299–317.

Beatty, John. 1994. "Dobzhansky and the Biology of Democracy: The Moral and Political Significance of Genetic Variation." In *The Evolution of Theodosius Dobzhansky: Essays on His Life and Thought in Russia and America*, edited by M. B. Adams, 195–218. Princeton, NJ: Princeton University Press.

Beckwith, Jon. 1991. "Foreword: The Human Genome Initiative: Genetics' Lightning Rod." *American Journal of Law and Medicine* 17: 1–13.

———. 2001. "On the Social Responsibility of Scientists." *Annali dell'Istituto Superiore di Sanità* 37: 189–194.

Beckwith, Jonathan. 2002. *Making Genes, Making Waves*. Cambridge, MA: Harvard University Press.

Beckwith, Jonathan, and Joseph S Alper. 1998. "Nature's Imperfect Experiment." *La Recherche* 311: 72–76.

Behrman, Jere R., and Paul Taubman. 1989. "Is Schooling 'Mostly in the Genes'? Nature-Nurture Decomposition Using Data on Relatives." *Journal of Political Economy* 97: 1425–46.

Benzer, S. 1971. "From the Gene to Behavior." *JAMA: The Journal of the American Medical Association* 218: 1015–22.

Bernhard, Robert. 1967. "Genetics and Human Intelligence." *Scientific Research* 2: 30–34.

BGA. 1985. "Members of the Behavior Genetics Association." *Behavior Genetics* 15: 189–96.

———. 1996. "Behavior Genetics Association Membership List, 1995–1996." *Behavior Genetics* 26: 335–62.

Billings, Paul R., Jonathan Beckwith, and Joseph S. Alper. 1992. "The Genetic Analysis of Human Behavior: A New Era?" *Social Science and Medicine* 35: 227–38.

Bliss, Catherine. 2012. *Race Decoded: The Genomic Fight for Social Justice*. Stanford, CA: Stanford University Press.

Bliss, E. L. 1962. *Roots of Behavior*. New York: Harper.

Block, N. J. 1996. "How Heritability Misleads about Race." *Boston Review* 20: 30–35.

Block, Ned, and Gerald Dworkin. 1974. "IQ: Heritability and Inequality." *Philosophy of Public Affairs* 3 (4): 331–409, 340–99.

Bloor, David. 1991. *Knowledge and Social Imagery*. Chicago: University of Chicago Press.

Blue, Laura. 2007. "The Mortification of James Watson." *Time*, October 19.

Blum, K., E. P. Noble, P. J. Sheridan, A. Montgomery, T. Ritchie, P. Jagadeeswaran, H. Nogami, A. H. Briggs, and J. B. Cohn. 1990. "Allelic Association of Human Dopamine D2 Receptor Gene in Alcoholism." *JAMA: The Journal of the American Medical Association* 263: 2055–60.

Blume, Stuart. 2006. "Anti-Vaccination Movements and Their Interpretations." *Social Science and Medicine* 62: 628–42.

Boas, Franz. 1940. *Race, Language, and Culture*. New York: Macmillan.

Bodmer, W. F., and L. L. Cavalli-Sforza. 1970. "Intelligence and Race." *Scientific American* 223: 19–29.

Bosk, Charles L. 2008. *What Would You Do? Juggling Bioethics and Ethnography*. Chicago: University of Chicago Press.

Bosker, F. J., C. A. Hartman, I. M. Nolte, B. P. Prins, P. Terpstra, D. Posthuma, T. van Veen, G. Willemsen, R. H. DeRijk, E. J. de Geus, W. J. Hoogendijk, P. F. Sullivan, B. W. Penninx, D. I. Boomsma, H. Snieder, and W. A. Nolen. 2011. "Poor Replication of Candidate Genes for Major Depressive Disorder Using Genome-Wide Association Data." *Molecular Psychiatry* 16: 516–32.

Botstein, David. 1997. "Of Genes and Genomes." In *Plain Talk about the Human Genome Project*, edited by E. Smith and W. Sapp, 207–14. Tuskegee, AL: Tuskegee University.

Bouchard, Thomas J. 1995. "Breaking the Last Taboo: Review of the Bell Curve." *Contemporary Psychology* 40: 415–18.

———. 2009. "Genetic Influence on Human Intelligence (Spearman's G): How Much?" *Annals of Human Biology* 36: 527–44.

Bouchard, Thomas J., Jr. 1994. "Genes, Environment, and Personality." *Science* 264: 1700–1701.

———. 1996. "Behaviour Genetic Studies of Intelligence, Yesterday and Today: The Long Journey from Plausibility to Proof." *Journal of Biosocial Science* 28: 527–55.

Bouchard, Thomas J., Jr., D. T. Lykken, Matthew McGue, Nancy L. Segal, and A. Tellegen. 1990. "Sources of Human Psychological Differences: The Minnesota Study of Twins Reared Apart." *Science* 250: 223–28.

Bouchard, Thomas J., Jr., and Matthew McGue. 1981. "Familial Studies of Intelligence: A Review." *Science* 212: 1055–59.

Bourdieu, Pierre. 1975. "The Specificity of the Scientific Field and the Social Conditions of the Progress of Reason." *Social Science Information / Information sur les Sciences Sociales* 14: 19–47.

———. 1988. *Homo Academicus*. Stanford, CA: Stanford University Press.

———. 1990. *The Logic of Practice*. Stanford, CA: Stanford University Press.

———. 1991. "The Peculiar History of Scientific Reason." *Sociological Forum* 6: 3–26.

———. 1993. *The Field of Cultural Production*. Translated by R. Johnson. New York: Columbia University Press.

———. 1996. *The Rules of Art: Genesis and Structure of the Literary Field*. Stanford, CA: Stanford University Press.

———. 1998. *Practical Reason: On the Theory of Action*. Stanford, CA: Stanford University Press.

———. 2004. *Science of Science and Reflexivity*. Translated by R. Nice. Chicago: University of Chicago Press.

Bourdieu, Pierre, and Loïc J. D. Wacquant. 1992. *An Invitation to Reflexive Sociology*. Chicago: University of Chicago Press.

Boyd, William C. 1950. *Genetics and the Races of Man*. Boston: D. C. Heath.

Brand, Christopher. 1999. "Genetic Science versus Authoritarianism on the Left: The Disruption of Yet Another Academic Meeting by Radical Protesters." *Mankind Quarterly* 40: 215–20.

Breggin, Peter R. 1995. "Campaigns against Racist Federal Programs by the Center for the Study of Psychiatry and Psychology." *Journal of African American Men* 1: 3–22.

Brimelow, P. 1994. "For Whom the Bell Tolls." *Forbes*: 153–63.

Brown, Nik, and Mike Michael. 2003. "A Sociology of Expectations: Retrospecting Prospects and Prospecting Retrospects." *Technology Analysis and Strategic Management* 15: 3.

Brown, Ryan P., and Robert A. Josephs. 1999. "A Burden of Proof: Stereotype Relevance and Gender Differences in Math Performance." *Journal of Personality and Social Psychology* 76: 246–57.

Bucchi, Massimiano. 1996. "When Scientists Turn to the Public: Alternative Routes in Science Communication." *Public Understanding of Science* 5: 375–94.

———. 1998. *Science and the Media: Alternative Routes in Scientific Communication*. London: Routledge.

Burkhardt, Richard W., Jr. 1981. "On the Emergence of Ethology as a Scientific Discipline." In *Conspectus of History*, vol. 1, edited by D. W. Hoover and J. T. A. Koumoulides. Muncie, IN: Ball State University Press.

Burr, Chandler. 1996. *A Separate Creation*. New York: Hyperion.

Butler, Declan. 1995. "Geneticist Quits in Protest at 'Genes and Violence' Claim." *Nature* 378: 224.

Buxbaum, J. D., S. Baron-Cohen, and B. Devlin. 2010. "Genetics in Psychiatry: Common Variant Association Studies." *Molecular Autism* 1: 6.

Calhoun, Craig. 1993. "Habitus, Field, and Capital: The Question of Historical Specificity." In *Bourdieu: Critical Perspectives*, edited by C. Calhoun, E. LiPuma, and M. Postone, 61–88. Chicago: University of Chicago Press.

Cambrosio, Alberto, and Peter Keating. 1983. "The Disciplinary Stake: The Case of Chronobiology." *Social Studies of Science* 13: 323–53.

Camic, Charles. 2006. "Book Review: Science of Science and Reflexivity." *American Journal of Sociology* 111: 1569–71.

Carlson, E. A. 2001. *The Unfit: A History of a Bad Idea*. Cold Spring Harbor, NY: Cold Spring Harbor Laboratory Press.

Carson, John. 2004. "The Science of Merit and the Merit of Science: Mental Order and Social Order in Early Twentieth-Century France and America." In *States of Knowledge: The Co-Production of Science and Social Order*, edited by S. Jasanoff, 181–205. London: Routledge.

———. 2007. *The Measure of Merit: Talents, Intelligence, and Inequality in the French and American Republics, 1750–1940.* Princeton, NJ: Princeton University Press.

Carson, Ronald A., and Mark A. Rothstein. 1999. *Behavioral Genetics: The Clash of Culture and Biology.* Baltimore: Johns Hopkins University Press.

Caspi, Avshalom, J. McClay, Terrie E. Moffitt, Jonathan Mill, Judy Martin, Ian W. Craig, Alan Taylor, and Richie Poulton. 2002. "Role of Genotype in the Cycle of Violence in Maltreated Children." *Science* 297: 851–54.

Caspi, Avshalom, Karen Sugden, Terrie E. Moffitt, Alan Taylor, Ian W. Craig, HonaLee Harrington, J. McClay, Jonathan Mill, Judy Martin, Antony Braithwaite, and Richie Poulton. 2003. "Influence of Life Stress on Depression: Moderation by a Polymorphism in the 5-Htt Gene." *Science* 301: 386–89.

Cassidy, Angela. 2005. "Popular Evolutionary Psychology in the UK: An Unusual Case of Science in the Media?" *Public Understanding of Science* 14: 115–41.

———. 2006. "Evolutionary Psychology as Public Science and Boundary Work." *Public Understanding of Science* 15: 175–205.

Cattell, Raymond B. 1980. "Cyril Burt, Psychologist." *Behavior Genetics* 10: 317–25.

Cavalli-Sforza, L. L., and Marcus W. Feldman. 1973. "Cultural versus Biological Inheritance: Phenotypic Transmission from Parents to Children (a Theory of the Effect of Parental Phenotypes on Children's Phenotypes)." *American Journal of Human Genetics* 25: 618–37.

Chabris, Christopher F., Benjamin M. Hebert, Daniel J. Benjamin, Jonathan P. Beauchamp, David Cesarini, Matthijs J. H. M. van der Loos, Magnus Johannesson, Patrik K. E. Magnusson, Paul Lichtenstein, Craig S. Atwood, Jeremy Freese, Taissa S. Hauser, Robert M. Hauser, Nicholas A. Christakis, and David Laibson. Forthcoming. "Most Reported Genetic Associations with General Intelligence Are Probably False Positives." *Psychological Science.*

Chang, Alicia, and Malcolm Ritter. 2011. "Specific IQ Genes Still Elusive, Latest Hunt Finds." *Seattle Times*, August 9.

Charney, Evan. 2012. "Behavior Genetics and Postgenomics." *Behavioral and Brain Sciences* 35: 1–80.

Charney, Evan, and William English. 2012. "Candidate Genes and Political Behavior." *American Political Science Review* 106: 1–34.

Chase, Allen. 1980. *The Legacy of Malthus.* Urbana: University of Illinois Press.

Chimonas, Susan, Lisa Patterson, Victoria H. Raveis, and David J. Rothman. 2011. "Managing Conflicts of Interest in Clinical Care: A National Survey of Policies at US Medical Schools." *Academic Medicine* 86: 293–99.

Clark, William R., and William Grunstein. 2000. *Are We Hardwired?* Oxford: Oxford University Press.

Clarke, Adele E. 1998. *Disciplining Reproduction: Modernity, American Life Sciences, and the Problems of Sex.* Berkeley: University of California Press.

Cloninger, C. R. 1986. "A Unified Biosocial Theory of Personality and Its Role in the Development of Anxiety States." *Psychiatric Developments* 4: 167–226.

Cloninger, C. Robert, John Rice, and Theodore Reich. 1979. "Multifactorial Inheritance with Cultural Transmission and Assortative Mating. II. A General Model of Combined Polygenic and Cultural Inheritance." *American Journal of Human Genetics* 31: 176–98.

Collins, H. M. 1981. "Knowledge and Controversy: Studies of Modern Natural Science." *Social Studies of Science* 11: 3–158.

———. 1985. *Changing Order: Replication and Induction in Scientific Practice.* London; Beverly Hills: Sage Publications.

———. 2000. "Surviving Closure: Post-Rejection Adaptation and Plurality in Science." *American Sociological Review* 65: 824–45.

Collins, Randall. 1998. *The Sociology of Philosophies: A Global Theory of Intellectual Change.* Cambridge, MA: Harvard University Press.

Condit, Celeste Michelle. 1999. *The Meanings of the Gene: Public Debates about Human Heredity.* Madison: University of Wisconsin Press.

Conrad, Peter. 1997. "Public Eyes and Private Genes: Historical Frames, News Constructions, and Social Problems." *Social Problems* 44: 139–54.

———. 2001. "Genetic Optimism: Framing Genes and Mental Illness in the News." *Culture, Medicine, and Psychiatry* 25: 225–47.

Conrad, Peter, and Dana Weinberg. 1996. "Has the Gene for Alcoholism Been Discovered Three Times since 1980? A News Media Analysis." *Perspectives on Social Problems* 8: 3–25.

Cooper, R. M., and J. P. Zubek. 1958. "Effects of Enriched and Restricted Early Environments on the Learning Ability of Rats." *Canadian Journal of Psychology* 12: 159–64.

Couldry, Nick. 2003. "Media Meta-Capital: Extending the Range of Bourdieu's Field Theory." *Theory and Society* 32: 653–77.

Council of the Academy. 1968. "A Statement by the Council of the Academy." *Proceedings of the National Academy of the Sciences* 59: 651–54.

Crabbe, John C., Douglas Wahlsten, and Bruce C. Dudek. 1999. "Genetics of Mouse Behavior: Interactions with Laboratory Environment." *Science* 284: 1670–72.

Crow, James F. 1969. "Genetic Theories and Influences: Comments on the Value of Diversity." *Harvard Educational Review* 39: 301–9.

Curtis, Bruce. 2003. "Book Reviews: Pierre Bourdieu, *Science de la science et réflexivité*." *Science, Technology, and Human Values* 28: 538–49.

Davies, G., A. Tenesa, A. Payton, J. Yang, S. E. Harris, D. Liewald, X. Ke, S. Le Hellard, A. Christoforou, M. Luciano, K. McGhee, L. Lopez, A. J. Gow, J. Corley, P. Redmond, H. C. Fox, P. Haggarty, L. J. Whalley, G. McNeill,

M. E. Goddard, T. Espeseth, A. J. Lundervold, I. Reinvang, A. Pickles, V. M. Steen, W. Ollier, D. J. Porteous, M. Horan, J. M. Starr, N. Pendleton, P. M. Visscher, and I. J. Deary. 2011. "Genome-Wide Association Studies Establish That Human Intelligence Is Highly Heritable and Polygenic." *Molecular Psychiatry* 16: 996–1005.

Davis, Bernard. 1975. "Social Determinism and Behavioral Genetics." *Science* 26: 189.

Dawkins, Richard. 1976. *The Selfish Gene*. Oxford: Oxford University Press.

Deary, Ian J., W. Johnson, and L. M. Houlihan. 2009. "Genetic Foundations of Human Intelligence." *Human Genetics* 126: 215–32.

DeCastro, John M. 1993. "A Twin Study of Genetic and Environmental Influences on the Intake of Fluids and Beverages." *Physiology and Behavior* 54: 677–87.

DeFries, J. C. 1972. "Reply to Professor Jensen." In *Genetics, Environment, and Behavior: Implications for Educational Policy*, edited by L. Ehrman, G. S. Omenn, and E. Caspari, 24–25. New York: Academic Press.

Detterman, Douglas K., Lee Anne Thompson, and Robert Plomin. 1990. "Differences in Heritability across Groups Differing in Ability." *Behavior Genetics* 20: 369–84.

Deutsch, Martin. 1969. "Happenings on the Way Back to the Forum: Social Science, IQ and Race Differences Revisited." *Harvard Educational Review* 39.

Devlin, Bernie, Stephen E. Feinberg, Daniel P. Resnick, and Kathryn Roeder. 1997. "Intelligence, Genes, and Success: Scientists Respond to *The Bell Curve*." New York: Springer.

Dewsbury, Donald A. 2009. "Origins of Behavior Genetics: The Role of the Jackson Laboratory." *Behavior Genetics* 39: 1–5.

Dick, D. M., R. J. Rose, R. J. Viken, J. Kaprio, and M. Koskenvuo. 2001. "Exploring Gene-Environment Interactions: Socioregional Moderation of Alcohol Use." *Journal of Abnormal Psychology* 110: 625–32.

Dick, Danielle M., and Richard J. Rose. 2002. "Behavior Genetics: What's New? What's Next?" *Current Directions in Psychological Science* 11: 70–74.

Dickens, William T., and James R. Flynn. 2001. "Heritability Estimates versus Large Environmental Effects: The IQ Paradox Resolved." *Psychological Review* 108: 346–69.

Dobzhansky, Theodosius. 1962. *Mankind Evolving: The Evolution of the Human Species*. New Haven, CT: Yale University Press.

——. 1968. "On Genetics, Sociology, and Politics." *Perspectives in Biology and Medicine* 11: 544–54.

——. 1973a. "Differences Are Not Deficits." *Psychology Today* 7: 97–101.

——. 1973b. *Genetic Diversity and Human Equality*. New York: Basic Books.

Duke, David. 1998. *My Awakening: A Path to Racial Understanding*. Covington, LA: Free Speech Press.

Dunn, L. C., and Theodosius Dobzhansky. 1952. *Heredity, Race, and Society.*
New York: New American Library.

Durkheim, Émile. 1951. *Suicide.* Translated by J. A. Spaulding and G. Simpson.
New York: Free Press.

Dusek, Val. 1987. "Bewitching Science: Twin Studies as Public Relations." *Science for the People* 19: 19–22.

Duster, Troy. 1990. *Backdoor to Eugenics.* New York: Routledge.

———. 2004. "Selective Arrests, an Ever-Expanding DNA Forensic Database,
and the Specter of an Early Twenty-First-Century Equivalent of Phrenology." In *The Technology of Justice: DNA and the Criminal Justice System,*
edited by D. Lazer, 315–334. Cambridge, MA: MIT Press.

———. 2006. "Behavioral Genetics and Explanations of the Link between Crime,
Violence, and Race." In *Wrestling with Behavioral Genetics: Science, Ethics,
and Public Conversation,* edited by E. Parens, A. R. Chapman and N. Press,
150–75. Baltimore: Johns Hopkins University Press.

Ebstein, Richard P., Olga Novick, Roberto Umansky, Beatrice Priel, Yamima
Osher, Darren Blaine, Estelle R. Bennett, Lubov Nemanov, Miri Katz, and
Robert H. Belmaker. 1996. "Dopamine D4 Receptor (D4dr) Exon III Polymorphism Associated with the Human Personality Trait of Novelty Seeking."
Nature Genetics 12: 78–80.

Eckert, E. D., T. J. Bouchard, J. Bohlen, and L. L. Heston. 1986. "Homosexuality in Monozygotic Twins Reared Apart." *British Journal of Psychiatry* 148:
421–25.

Egeland, Janice A., Daniela S. Gerhard, David L. Pauls, James N. Sussex, Kenneth K. Kidd, Cleona R. Alien, Abram M. Hostetter, and David E. Housman. 1987. "Bipolar Affective Disorders Linked to DNA Markers on Chromosome 11." *Nature* 325: 783–87.

Ehrman, Lee, Gilbert S. Omenn, and Ernst Caspari. 1972. *Genetics, Environment, and Behavior: Implications for Educational Policy.* New York: Academic Press.

Elias, Norbert, and Eric Dunning. 1994. *Quest for Excitement: Sport and Leisure in the Civilizing Process.* London: Blackwell.

Enserink, Martin. 1999. "Fickle Mice Highlight Test Problems." *Science* 284:
1599–1600.

Epstein, Steven. 1996. *Impure Science: Aids, Activism, and the Politics of Knowledge.* Berkeley: University of California Press.

———. 2007. *Inclusion: The Politics of Difference in Medical Research.* Chicago:
University of Chicago Press.

Erlenmeyer-Kimling, L., and Lissy F. Jarvik. 1963. "Genetics and Intelligence:
A Review." *Science* 142: 1477–79.

Evans, John H. 2002. *Playing God? Human Genetic Engineering and the Ra-*

tionalization of Public Bioethical Debate. Chicago: University of Chicago Press.

———. 2012. *The History and Future of Bioethics: A Sociological View.* New York: Oxford University Press.

Evans, Patrick D., Sandra L. Gilbert, Nitzan Mekel-Bobrov, Eric J. Vallender, Jeffrey R. Anderson, Leila M. Vaez-Azizi, Sarah A. Tishkoff, Richard R. Hudson, and Bruce T. Lahn. 2005. "Microcephalin, a Gene Regulating Brain Size, Continues to Evolve Adaptively in Humans." *Science* 309: 1717–20.

Eyal, Gil. 2002. "Dangerous Liaisons between Military Intelligence and Middle Eastern Studies in Israel." *Theory and Society* 31: 653–93.

Eysenck, Hans Jürgen. 1971. *The IQ Argument: Race, Intelligence, and Education.* London: Temple Smith.

Ezrahi, Yaron. 1976. "The Jensen Controversy: A Study in the Ethics and Politics of Knowledge in Democracy." In *Controversies and Decisions: The Social Sciences and Public Policy,* edited by C. Frankel, 149–70. New York: Russell Sage Foundation.

Farahany, Nita, and William Bernet. 2006. "Behavioural Genetics in Criminal Cases: Past, Present, and Future." *Genomics, Society, and Policy* 2: 72–79.

Farrell, Michael B. 2010. "Prop. 8 Trial: Defenders of Gay-Marriage Ban Make Their Case." *Christian Science Monitor.*

Feldman, Marcus W., and Richard C. Lewontin. 1975. "The Heritability 'Hang-Up.'" *Science* 190: 1163–68.

Feresin, Emiliano. 2009. "Lighter Sentence for Murderer with 'Bad Genes.'" *Nature.* doi:10.1038/news.2009.1050.

Fischer, Claude S., Michael Hout, Martin Sanchez Jankowski, Samuel R. Lucas, Ann Swidler, and Kim Voss. 1996. *Inequality by Design: Cracking the Bell Curve Myth.* Princeton, NJ: Princeton University Press.

Fishbein, Diana H. 1996. "Prospects for the Application of Genetic Findings to Crime and Violence Prevention." *Politics and the Life Sciences* 15: 91–94.

Flaschman, Stu, and Alan Weintraub. 1974. "Heritability: A Scientific Snow Job." *Science for the People* 6: 21–22.

Fligstein, Neil, and Doug McAdam. 2012. *A Theory of Fields.* New York: Oxford University Press.

Fox, Robin. 1976. "On the Genetics of Being Human." In *Human Behavior Genetics,* edited by A. R. Kaplan, 49–62. Springfield, IL: Charles C. Thomas.

Fraser, Steven. 1995. *The Bell Curve Wars: Race, Intelligence, and the Future of America.* New York: Basic Books.

Fred Friendly Seminars, Inc. 2002. "Genes on Trial." PBS, *Our Genes / Our Choices.*

Freedman, Daniel G. 1965. "An Ethological Approach to the Genetical Study of Human Behavior." In *Methods and Goals in Human Behavior Genetics,* edited by S. G. Vandenberg, 141–62. New York: Academic Press.

———. 1968. "An Evolutionary Framework for Behavioral Research." In *Progress in Human Behavior Genetics*, edited by S. G. Vandenberg, 1–6. Baltimore: Johns Hopkins University Press.

Frickel, Scott. 2004a. "Building an Interdiscipline: Collective Action Framing and the Rise of Genetic Toxicology." *Social Problems* 51: 269–87.

———. 2004b. *Chemical Consequences: Environmental Mutagens, Scientist Activism, and the Rise of Genetic Toxicology*. New Brunswick, NJ: Rutgers University Press.

———. 2006. "When Convention Becomes Contentious: Organizing Science Activism in Genetic Toxicology." In *The New Political Sociology of Science*, edited by S. Frickel and K. Moore, 185–214. Madison: University of Wisconsin Press.

Frickel, Scott, Sahra Gibbon, Jeff Howard, Joanna Kempner, Gwen Ottinger, and David J. Hess. 2010. "Undone Science: Charting Social Movement and Civil Society Challenges to Research Agenda Setting." *Science, Technology, and Human Values* 35: 444–73.

Frickel, Scott, and Neil Gross. 2005. "A General Theory of Scientific/Intellectual Movements." *American Sociological Review* 70: 204–32.

Friedrichs, Robert W. 1973. "The Impact of Social Factors upon Scientific Judgment: The 'Jensen Thesis' as Appraised by Members of the American Psychological Association." *Journal of Negro Education* 42: 429–38.

Fujimura, Joan H. 1996. *Crafting Science: A Sociohistory of the Quest for the Genetics of Cancer*. Cambridge, MA: Harvard University Press.

Fukuyama, Francis. 2002. *Our Posthuman Future: Consequences of the Biotechnology Revolution*. New York: Picador.

Fuller, John L. 1960. "Behavior Genetics." *Annual Review of Psychology* 11: 41–70.

———. 1983. "Sociobiology and Behavior Genetics." In *Behavior Genetics: Principles and Applications*, edited by J. L. Fuller and E. C. Simmel. Hillsdale, NJ: Lawrence Erlbaum Associates.

———. 1985. "Of Dogs, Mice, People, and Me." In *Leaders in the Study of Animal Behavior*, edited by D. A. Dewsbury, 92–118. London: Associated University Presses.

Fuller, John L., and W. Robert Thompson. 1960. *Behavior Genetics*. New York: Wiley.

Gadamer, Hans-Georg. 2004. *Truth and Method*. London: Continuum.

Gaylin, William. 1980. "The XYY Controversy: Researching Violence and Genetics." *Hastings Center Report, Special Supplement*: 1–32.

Gibbons, Michael, Camille Limoges, Helga Nowotny, Simon Schwartzman, Peter Scott, and Martin Trow. 1994. *The New Production of Knowledge*. New York: Sage.

Giddens, Anthony. 1984. *The Constitution of Society*. Cambridge: Polity.

Gieryn, Thomas F. 1983. "Boundary-Work and the Demarcation of Science from Non-Science: Strains and Interests in Professional Ideologies of Scientists." *American Sociological Review* 48: 781–95.

———. 1999. *Cultural Boundaries of Science: Credibility on the Line.* Chicago: University of Chicago Press.

Gilbert, Walter. 1992. "A Vision of the Grail." In *The Code of Codes: Scientific and Social Issues in the Human Genome Project,* edited by D. J. Kevles and L. E. Hood, 83–97. Cambridge, MA: Harvard University Press.

Ginsburg, Benson E., and William S. Laughlin. 1966. "The Multiple Bases of Human Adaptability and Achievement: A Species Point of View." *Eugenics Quarterly* 13: 240–57.

Glass, David G. 1968. *Biology and Behavior: Genetics.* New York: Rockefeller University Press.

Goldberger, Arthur S. 1976. "Mysteries of the Meritocracy." In *The IQ Controversy,* edited by N. J. Block and G. Dworkin, 265–79. New York: Pantheon.

———. 1979. "Heritability." *Economica* 46: 327–47.

Goldberger, Arthur S., and Nicholas M. Kiefer. 1989. "The ET Interview: Arthur S. Goldberger." *Econometric Theory* 5: 133–60.

Goldenberg, Suzanne. 2012. "Climate Scientist Peter Gleick Admits He Leaked Heartland Institute Documents." *The* (Manchester) *Guardian.*

Gorroochurn, Prakash, Susan E. Hodge, Gary A. Heiman, Martina Durner, and David A. Greenberg. 2007. "Non-Replication of Association Studies: 'Pseudo-Failures' To Replicate?" *Genetics in Medicine* 9: 325–31.

Gottesman, Irving I. 1965. "Personality and Natural Selection." In *Methods and Goals in Human Behavior Genetics,* edited by S. G. Vandenberg, 63–80. New York: Academic Press.

———. 1974. "Exorcise or Excommunicate the IQ-Race-Class Controversy, Review of Dobzhansky, Genetic Diversity and Human Equality." *Contemporary Psychology* 19: 587–88.

———. 2008. "Milestones in the History of Behavioral Genetics: Participant Observer." *Acta Psychologica Sinica* 40: 1042–50.

Gottesman, Irving I., and L. Erlenmeyer-Kimling. 1971. "A Foundation for Informed Eugenics." *Social Biology* 18: S1–S8.

Gottesman, Irving I., and Todd D. Gould. 2003. "The Endophenotype Concept in Psychiatry: Etymology and Strategic Intentions." *American Journal of Psychiatry* 106: 636–45.

Gottesman, Irving I., and James Shields. 1967. "A Polygenic Theory of Schizophrenia." *Proceedings of the National Academy of Sciences* 58: 199–205.

———. 1972. *Schizophrenia and Genetics: A Twin Study Vantage Point.* New York: Academic Press.

———. 1976. "A Critical Review of Recent Adoption, Twin, and Family Studies of Schizophrenia: Behavioral Genetics Perspectives." *Schizophrenia Bulletin* 2: 360–401.

Gottfredson, Linda S. 1994. "Egalitarian Fiction and Collective Fraud." *Society* 31: 53–59.

———. 1997. "Mainstream Science on Intelligence: An Editorial with 52 Signatories, History, and Bibliography." *Intelligence* 24: 13–23.

———. 2010. "Lessons in Academic Freedom as Lived Experience." *Personality and Individual Differences* 49: 272–80.

Gottlieb, G. 1995. "Some Conceptual Deficiencies in 'Developmental' Behavior Genetics." *Human Development* 38: 131–41.

Gould, Stephen Jay. 1996. *The Mismeasure of Man.* New York: Norton.

Green, Jeremy. 1985. "Media Sensationalism and Science: The Case of the Criminal Chromosome." In *Expository Science*, Sociology of the Sciences 9, edited by T. Shinn and R. Whitley, 139–61. Boston: Kluwer Academic.

Greenberg, Daniel. 2001. *Science, Money, and Politics.* Chicago: University of Chicago Press.

Griffiths, Paul E., and James Tabery. 2008. "Behavioral Genetics and Development: Historical and Conceptual Causes of Controversy." *New Ideas in Psychology* 26: 332–52.

Habermas, Jürgen. 2003. *The Future of Human Nature.* Cambridge: Polity.

Hacking, Ian. 1999. *The Social Construction of What?* Cambridge, MA: Harvard University Press.

Hall, Calvin S. 1951. "The Genetics of Behavior." In *Handbook of Experimental Psychology*, edited by S. S. Stevens, 304–29. New York: John Wiley and Sons.

Hamer, Dean H. 2004. *The God Gene.* New York: Doubleday.

Hamer, Dean H., and Peter Copeland. 1994. *The Science of Desire: The Search for the Gay Gene and the Biology of Behavior.* New York: Simon and Schuster.

———. 1998. *Living with Our Genes: Why They Matter More Than You Think.* New York: Doubleday.

Hamer, Dean H., Stella Hu, Victoria L. Magnuson, Nan Hu, and Angela M. L. Pattatucci. 1993. "A Linkage between DNA Markers on the X Chromosome and Male Sexual Orientation." *Science* 261: 321–25.

Harris, Judith Rich. 1998. *The Nurture Assumption: Why Children Turn Out the Way They Do.* New York: Free Press.

Harvey, David. 2005. *A Brief History of Neoliberalism.* Oxford: Oxford University Press.

Harwood, Jonathan. 1976. "The Race-Intelligence Controversy: A Sociological Approach. I—Professional Factors." *Social Studies of Science* 6: 369–94.

———. 1977. "The Race-Intelligence Controversy: A Sociological Approach. II—'External' Factors." *Social Studies of Science* 7: 1–30.

———. 1979. "Heredity, Environment, and the Legitimation of Social Policy." In *Natural Order: Historical Studies of Scientific Culture*, edited by B. Barnes and S. Shapin, 231–48. Beverly Hills: Sage.

Haworth, Claire M. A., and Robert Plomin. 2010. "Quantitative Genetics in the Era of Molecular Genetics: Learning Abilities and Disabilities as an Example." *Journal of the American Academy of Child and Adolescent Psychiatry* 49: 783–93.

Hay, D. A. 2003. "Who Should Fund and Control the Direction of Human Behavior Genetics?" *Genes, Brain, and Behavior* 2: 321–26.

Hay, David A. 1985. *Essentials of Behavior Genetics*. Melbourne: Blackwell Scientific.

Heath, Andrew C. 1995. "The 25th Annual Meeting of the Behavior Genetics Association, Richmond, Virginia." *Behavior Genetics* 26: 589–90.

Hedgecoe, Adam. 1998. "Geneticization, Medicalisation, and Polemics." *Medicine, Health Care, and Philosophy* 1: 235–43.

Helmreich, Stefan. 2003. "Trees and Seas of Information: Alien Kinship and the Biopolitics of Gene Transfer in Marine Biology and Biotechnology." *American Ethnologist* 30: 340–58.

Herrnstein, Richard J. 1971. "IQ." *Atlantic Monthly* 228: 43–64.

———. 1973. *I.Q. in the Meritocracy*. Boston: Little, Brown.

Herrnstein, Richard J., and Charles A. Murray. 1994. *The Bell Curve: Intelligence and Class Structure in American Life*. New York: Free Press.

Hess, David J. 2004. "Medical Modernisation, Scientific Research Fields, and the Epistemic Politics of Health Social Movements." *Sociology of Health and Illness* 26: 695–709.

Higgins, J. V., Elizabeth W. Reed, and S. C. Reed. 1962. "Intelligence and Family Size: A Paradox Resolved." *Eugenics Quarterly* 9: 84–90.

Hilgartner, Stephen. 2000. *Science on Stage*. Stanford, CA: Stanford University Press.

Hirsch, Jerry. 1962. "Individual Differences in Behavior and Their Genetic Basis." In *Roots of Behavior*, edited by E. Bliss, 3–23. New York: Paul B. Hoeber.

———. 1963. "Behavior Genetics and Individuality Understood: Behaviorism's Counterfactual Dogma Blinded the Behavioral Sciences to the Significance of Meiosis." *Science* 142: 1436–42.

———. 1967a. *Behavior-Genetic Analysis*. New York: McGraw-Hill.

———. 1967b. "Behavior-Genetic, or 'Experimental,' Analysis: The Challenge of Science versus the Lure of Technology." *American Psychologist* 22: 118–30.

———. 1967c. "Epilog: Behavior-Genetic Analysis." In *Behavior-Genetic Analysis*, edited by J. Hirsch. New York: McGraw-Hill.

———. 1967d. Preface to *Behavior-Genetic Analysis*, edited by J. Hirsch, xi–xii. New York: McGraw-Hill.

———. 1968. "Behavior-Genetic Analysis and the Study of Man." In *Science and the Concept of Race*, edited by Margaret Mead and others, 37–48. New York: Columbia University Press.

———. 1970. "Behavior-Genetic Analysis and Its Biosocial Consequences." *Seminars in Psychiatry* 2: 89–105.

———. 1975. "Jensenism: The Bankruptcy of 'Science' without Scholarship." *Educational Theory* 25: 3–27, 102.

———. 1978. "Evidence for Equality: Genetic Diversity and Social Organization." In *Equality and Social Policy*, edited by W. Feinberg, 143–62. Urbana: University of Illinois Press.

———. 1981. "To 'Unfrock the Charlatans.'" *Sage Race Relations Abstracts* 6: 1–67.

———. 1997a. "Introduction." *Genetica* 99: 75–76.

———. 1997b. "Some History of Heredity-vs-Environment, Genetic Inferiority at Harvard(?), and *The* (Incredible) *Bell Curve*." *Genetica* 99: 207–24.

Hirsch, Jerry, M. Beeman, and Tim Tully. 1980a. "Compensatory Education Has Succeeded [Review of A. R. Jensen (1980), *Bias in Mental Testing*]." *Behavioral and Brain Sciences* 3.

Hirsch, Jerry, Terry R. McGuire, and Atam Vetta. 1980b. "Concepts of Behavior Genetics and Misapplications to Humans." In *The Evolution of Human Social Behavior*, edited by J. Lockard, 215–38. New York: North-Holland.

Hirsch, Jerry, and Tim Tully. 1982. "The Challenge Is Unmet [Continuing Commentary on A. R. Jensen 1980]." *Behavioral and Brain Sciences* 5: 324–26.

Hirschhorn, Joel N., Kirk Lohmuller, Edward Byrne, and Kurt Hirschhorn. 2002. "A Comprehensive Review of Genetic Association Studies." *Genetics in Medicine* 4: 45–61.

Hirschman, Albert O. 1970. *Exit, Voice, and Loyalty*. Cambridge, MA: Harvard University Press.

———. 1991. *The Rhetoric of Reaction: Perversity, Futility, Jeopardy*. Cambridge, MA: Belknap Press.

Holden, C. 1980. "Identical Twins Reared Apart." *Science* 207: 1323–25, 1327–28.

———. 1995a. "Behavior Geneticists Shun Colleague." *Science* 270: 1125.

———. 1995b. "Specter at the Feast." *Science* 269: 35.

Holden, C., and T. Bouchard. 2009. "Behavioral Geneticist Celebrates Twins, Scorns PC Science." *Science* 325: 27–27.

Holden, Constance. 1994. "A Cautionary Genetic Tale: The Sobering Story of D2." *Science* 264: 1696–97.

Horn, J. M. 1983. "The Texas Adoption Project: Adopted Children and Their Intellectual Resemblance to Biological and Adoptive Parents." *Child Development* 54: 268–75.

Horn, J. M., J. C. Loehlin, and L. Willerman. 1979. "Intellectual Resemblance among Adoptive and Biological Relatives: The Texas Adoption Project." *Behavior Genetics* 9: 177–207.

Horowitz, Irving L. 1995. "The Rushton File." In *The Bell Curve Debate*, edited by R. Jacoby and N. Glauberman, 179–200. New York: Times Books.

———. 2000. "Letter 'to Our Friends, Readers, and Subscribers.'" *Society* 36: front cover.

Horwitz, Allan V., Tami M. Videon, Mark F. Schmitz, and Diane Davis. 2003. "Rethinking Twins and Environments: Possible Social Sources for Assumed Genetic Influences in Twin Research." *Journal of Health and Social Behavior* 44: 111–29.

Hubbard, Ruth, and Elijah Wald. 1993. *Exploding the Gene Myth*. Boston: Beacon Press.

Hughes, Robert. 1991. *The Shock of the New*. New York: Knopf.

Hunt, Earl. 2010. "The Rights and Responsibilities Implied by Academic Freedom." *Personality and Individual Differences* 49: 264–71.

Hunt, Morton. 1999. *The New Know-Nothings: The Political Foes of the Scientific Study of Human Nature*. New Brunswick, NJ: Transaction Publishers.

Ioannidis, John P. A. 2005. "Why Most Published Research Findings Are False." *PLOS Medicine* 2: e124.

Jackson, J. F. 1993. "Human Behavioral Genetics, Scarr's Theory, and Her Views on Interventions: A Critical Review and Commentary on Their Implications for African American Children." *Child Development* 64: 1318–32.

Jacobs, Jerry A., and Scott Frickel. 2009. "Interdisciplinarity: A Critical Assessment." *Annual Review of Sociology* 35: 43–65.

Jacobs, P. A., M. Brunton, M. M. Melville, R. P. Brittain, and W. F. McClemont. 1965. "Aggressive Behavior, Mental Subnormality, and the XYY Male." *Nature* 208: 1351–52.

Jacoby, Russell, and Naomi Glauberman, eds. 1995. *The Bell Curve Debate: History, Documents, Opinions*. New York: Times Books, xiv, 720.

Jaroff, L. 1989. "The Gene Hunt." *Time*, March 20, 62–67.

Jasanoff, Sheila, ed. 2004. *States of Knowledge: The Co-Production of Science and Social Order*. London: Routledge.

Jasper, James M., and Dorothy Nelkin. 1992. *The Animal Rights Crusade: The Growth of a Moral Protest*. New York: Free Press.

Jencks, Christopher. 1980. "Heredity, Environment, and Public Policy Reconsidered." *American Sociological Review* 45: 723–36.

Jencks, Christopher, Marshall Smith, Henrey Ackland, Mary Jo Bane, David Cohen, Herbert Gintis, Barbara Heyns, and Stephan Michelson. 1973. *Inequality*. New York: Harper Colophon.

Jensen, Arthur R. 1967. "The Culturally Disadvantaged: Psychological and Educational Aspects." *Educational Research* 10.

———. 1969. "How Much Can We Boost IQ and Scholastic Achievement?" *Harvard Educational Review* 39: 1–123.

———. 1972. Preface to *Genetics and Education*, edited by A. R. Jensen, 1–67. New York: Harper and Row.

———. 1973. *Educability and Group Differences*. New York: Harper and Row.

———. 1976. "Twins' IQs: A Reply to Schwartz and Schwartz." *Behavior Genetics* 6: 369–71.

———. 1980. "Correcting the Bias against Mental Testing: A Preponderance of Peer Agreement." *Behavioral and Brain Sciences* 3: 359–71.

Johnson, Wendy, and Matt McGue. 2010. "Tom Bouchard: A Compelling Presence in Psychology." *Personality and Individual Differences* 49: 261–63.

Joseph, Jay. 1998. "The Equal Environment Assumption of the Classical Twin Method: A Critical Analysis." *Journal of Mind and Behavior* 19: 325–58.

———. 2004. *The Gene Illusion: Genetic Research in Psychiatry and Psychology under the Microscope*. New York: Algora.

———. 2006. *The Missing Gene: Psychiatry, Heredity, and the Fruitless Search for Genes*. New York: Algora.

———. 2011. "The Crumbling Pillars of Behavioral Genetics." *GeneWatch* 24: 4–7.

Kagan, Jerome. 1969. "Inadequate Evidence and Illogical Conclusions." *Harvard Educational Review* 39: 274–77.

———. 1984. *The Nature of the Child*. New York: Basic Books.

———. 2003. "A Behavioral Science Perspective." In *Behavioral Genetics in the Postgenomic Era*, edited by R. Plomin, J. C. DeFries, I. W. Craig, and P. McGuffin, xvii–xx. Washington, DC: American Psychological Association.

Kamin, Leon J. 1974. *The Science and Politics of I.Q.* Potomac, MD: L. Erlbaum Associates.

———. 1977a. "Comment on Munsinger's Adoption Study." *Behavior Genetics* 7: 403–6.

———. 1977b. "A Reply to Munsinger." *Behavior Genetics* 7: 411–12.

———. 1978. "The Hawaii Family Study of Cognitive Abilities: A Comment." *Behavior Genetics* 8: 275–79.

———. 1981. "Commentary." In *Race, Social Class, and Individual Differences in IQ*, edited by S. Scarr. Hillsdale, NJ: Laurence Erlbaum Associates.

Kamin, Leon J., and Arthur S. Goldberger. 2002. "Twin Studies in Behavioral Research: A Skeptical View." *Theoretical Population Biology* 61: 83–95.

Kaplan, Jonathan. 2006. "Misinformation, Misrepresentation, and Misuse of Human Behavioral Genetics Research." *Law and Contemporary Problems* 69: 47–80.

Kaplan, Jonathan Michael. 2000. *The Limits and Lies of Human Genetic Research: Dangers for Social Policy*. New York: Routledge.

Kato, Tadafumi. 2007. "Molecular Genetics of Bipolar Disorder and Depression." *Psychiatry and Clinical Neurosciences* 61: 3–19.

Keller, Evelyn Fox. 2010. *The Mirage of a Space between Nature and Nurture.* Durham, NC: Duke University Press.

Kelsoe, John R., Edward I. Ginns, Janice A. Egeland, Daniela S. Gerhard, Alisa M. Goldstein, Sherri J. Bale, David L. Pauls, Robert T. Long, Kenneth K. Kidd, Giovanni Conte, David E. Housman, and Steven M. Paul. 1989. "Re-Evaluation of the Linkage Relationship between Chromosome 11p Loci and the Gene for Bipolar Affective Disorder in the Old Order Amish." *Nature* 342: 238–43.

Kempthorne, Oscar. 1978. "Logical, Epistemological, and Statistical Aspects of Nature-Nurture Data Interpretation." *Biometrics* 34: 1–23.

Kendler, Kenneth S. 1994. "Discussion: Genetic Analysis." In *Genetic Approaches to Mental Disorders*, edited by E. S. Gershon and C. R. Cloninger. Washington, DC: American Psychiatric Press.

———. 2005. "Psychiatric Genetics: A Methodologic Critique." *American Journal of Psychiatry* 162: 3–11.

Kendler, Kenneth S., Michael C. Neale, Ronald C. Kessler, Andrew C. Heath, and Lindon J. Eaves. 1993. "A Test of the Equal-Environment Assumption in Twin Studies of Psychiatric Illness." *Behavior Genetics* 23: 21–27.

Kevles, Daniel J. 1985. *In the Name of Eugenics: Genetics and the Uses of Human Heredity.* New York: Knopf.

Kim, Kyung-Man. 2009. "What Would a Bourdieuian Sociology of Scientific Truth Look Like?" *Social Science Information* 48: 57–79.

Kincheloe, Joe L., Shirley R. Steinberg, and Aaron D. Gresson III. 1996. *Measured Lies: The Bell Curve Examined.* New York: St. Martin's Press.

Kirp, David. 2003. *Shakespeare, Einstein, and the Bottom Line: The Marketing of Higher Education.* Cambridge, MA: Harvard University Press.

———. 2007. *The Sandbox Investment: The Preschool Movement and Kids-First Politics.* Cambridge, MA: Harvard University Press.

Klein, Julie Thompson. 1996. *Crossing Boundaries: Knowledge, Disciplinarities, and Interdisciplinarities.* Charlottesville: University Press of Virginia.

Klineberg, Otto. 1963. "Negro-White Differences in Intelligence Test Performance: A New Look at an Old Problem." *American Psychologist:* 198–203.

Knorr-Cetina, K. 1999. *Epistemic Cultures: How the Sciences Make Knowledge.* Cambridge, MA: Harvard University Press.

Koshland, D. 1987. "Nature, Nurture, and Behavior." *Science* 235: 1445.

———. 1989. "Sequences and Consequences of the Human Genome." *Science* 246: 189.

———. 1990. "A Rational Approach to the Irrational." *Science* 250: 189.

Krimsky, Sheldon. 2003. *Science in the Private Interest.* Lanham, MD: Rowman and Littlefield.

Kuhn, Thomas S. 1970. *The Structure of Scientific Revolutions*. Chicago: University of Chicago Press.

Lakatos, Imre. 1970. "Falsification and the Methodology of Scientific Research Programs." In *Criticism and the Growth of Knowledge*, edited by I. Lakatos and A. Musgrave, 91–195. Cambridge: Cambridge University Press.

Lamont, Michèle. 1992. *Money, Morals, and Manners: The Culture of the French and the American Upper-Middle Class*. Chicago: University of Chicago Press.

———. 2009. *How Professors Think*. Cambridge, MA: Harvard University Press.

Larson, Christina. 2013. "Sequencing a Complete Human Genome May Soon Cost Less Than an iPhone: Will China's BGI-Shenzhen Decode Yours?" *MIT Technology Review* (March/April): 34–37.

Latour, Bruno. 1987. *Science in Action: How to Follow Scientists and Engineers through Society*. Cambridge, MA: Harvard University Press.

Latour, Bruno, and Steve Woolgar. 1986. *Laboratory Life: The Construction of Scientific Facts*. Princeton, NJ: Princeton University Press.

Layzer, D. 1974. "Heritability Analyses of I.Q. Scores: Science or Numerology?" *Science* 183: 1259–66.

Lederberg, Joshua. 1973. "The Genetics of Human Nature." *Social Research* 40: 375–406.

Lemaine, Gerard, Roy MacLeod, Michael Mulkay, and Peter Weingart. 1976. *Perspectives on the Emergence of Scientific Disciplines*. The Hague: Mouton.

Lerner, R. M. 2006. "Another Nine-Inch Nail for Behavioral Genetics!" *Human Development* 49: 336–42.

Levins, Richard, and Richard C. Lewontin. 1985. *The Dialectical Biologist*. Cambridge, MA: Harvard University Press.

Lewontin, Richard C. 1970a. "Further Remarks on Race and the Genetics of Intelligence." *Bulletin of the Atomic Scientists* 26: 23–25.

———. 1970b. "Race and Intelligence." *Bulletin of the Atomic Scientists* 26: 2–8.

———. 1972. "The Apportionment of Human Diversity." *Evolutionary Biology* 6: 381–98.

———. 1974. "The Analysis of Variance and the Analysis of Causes." *American Journal of Human Genetics* 26: 400–411.

———. 1975. "Genetic Aspects of Intelligence." *Annual Review of Genetics* 9: 387–405.

———. 1976a. "Comment on an Erroneous Conception of the Meaning of Heritability." *Behavior Genetics* 6: 373–74.

———. 1976b. "The Fallacy of Biological Determinism." *The Sciences*: 6–10.

———. 1976c. "Race Differences in Intelligence, Book Review." *American Journal of Human Genetics* 28: 92–97.

———. 1996. "Of Genes and Genitals." *Transition*: 178–93.

Lewontin, Richard C., and L. L. Cavalli-Sforza. 1974. "Summary of Workshop on Population Genetics." National Institutes of Health Genetics Study Section.

Lewontin, Richard C., Steven P. R. Rose, and Leon J. Kamin. 1984. *Not in Our Genes: Biology, Ideology, and Human Nature.* New York: Pantheon Books.

Lichtenstein, Paul, Nancy L. Pedersen, and G. E. McClearn. 1992. "The Origins of Individual Differences in Occupational Status and Educational Level." *Acta Sociologica* 35: 13–31.

Lieberman, Leonard. 1968. "The Debate over Race: A Study in the Sociology of Knowledge." *Phylon (1960–)* 29: 127–41.

Lieberman, Leonard, and Larry T. Reynolds. 1978. "The Debate over Race Revisited: An Empirical Investigation." *Phylon (1960–)* 39: 333–43.

Lindee, M. Susan. 2005. *Moments of Truth in Genetic Medicine.* Baltimore: Johns Hopkins University Press.

Lippman, Abby. 1991. "Prenatal Genetic Testing and Screening: Constructing Needs and Reinforcing Inequities." *American Journal of Law and Medicine* 17: 15–50.

———. 1992. "Led (Astray) by Genetic Maps: The Cartography of the Human Genome and Health Care." *Social Science and Medicine* 35: 1469–76.

Lite, Jordan 2009. "Do Our Genes Make Us Popular?" *Scientific American*, January 26.

Loehlin, J. C., G. Lindzey, and J. N. Spuhler. 1975. *Race Differences in Intelligence.* San Francisco: W. H. Freeman.

Loehlin, John C. 1992. "Editorial: Should We Do Research on Race Differences in Intelligence?" *Intelligence* 16: 1–4.

———. 2009. "History of Behavior Genetics." In *Handbook of Behavior Genetics*, edited by Y.-K. Kim, 3–11. New York: Springer.

Loehlin, John C., and Robert C. Nichols. 1976. *Heredity, Environment, and Personality.* Austin: University of Texas Press.

Lombardo, Paul A. 2002. "'The American Breed': Nazi Eugenics and the Origins of the Pioneer Fund." *Albany Law Review* 65: 743–829.

Ludmerer, K. 1972. *Genetics and American Society.* Baltimore: Johns Hopkins University Press.

Lykken, David T. 2002. "A Professional Autobiography: Do the Mistakes of Youth Become the Wisdom of Old Age?" Pp. 1–49. http://www.psych.umn.edu/faculty/lykken/autobio.pdf.

———. 2006. "A Professional Autobiography: Fortune Doesn't Matter, but Good Fortune Does." Pp. 1–61. home.fnal.gov/~lykken/Autobiography.pdf.

Lykken, David T., Matthew McGue, Thomas J. Bouchard, and Auke Tellegen. 1990. "Does Contact Lead to Similarity or Similarity to Contact?" *Behavior Genetics* 20: 547–61.

Lynch, Michael. 1985. *Art and Artifact in Laboratory Science: A Study of Shop*

Work and Shop Talk in a Research Laboratory. London; Boston: Routledge and Kegan Paul.

Lynn, Richard. 2001. *The Science of Human Diversity: A History of the Pioneer Fund.* Lanham, MD: University Press of America.

Lyons, Michael J., Jack Goldberg, Seth A. Eisen, William True, Ming T. Tsuang, Joanne M. Meyer, and William G. Henderson. 1993. "Do Genes Influence Exposure to Trauma? A Twin Study of Combat." *American Journal of Medical Genetics* 48: 22–27.

Maccoby, Eleanor E. 2000. "Parenting and Its Effects on Children: On Reading and Misreading Behavior Genetics." *Annual Review of Psychology* 51: 1–27.

Macrae, Fiona. 2009. "If You Are a Gambler, You Can Bet It Is in Your Genes." (London) *Daily Mail.*

Maher, Brendan. 2008. "Personal Genomes: The Case of the Missing Heritability." *Nature* 456: 18–21.

Manzi, Jim. 2008. "Undetermined." *National Review*: 26–32.

Marshall, Eliot. 1994. "Highs and Lows on the Research Roller Coaster." *Science* 264: 1693–95.

Martin, John Levi. 2003. "What Is Field Theory?" *American Journal of Sociology* 109: 1–49.

Martin, N. G., L. J. Eaves, A. C. Heath, R. Jardine, L. M. Feingold, and H. J. Eysenck. 1986. "Transmission of Social Attitudes." *Proceedings of the National Academy of Sciences* 83: 4364–68.

Masters, Roger D. 1996. "Neuroscience, Genetics, and Society: Is the Biology of Human Social Behavior Too Controversial to Study?" *Politics and the Life Sciences* 15: 103–4.

Maxson, Stephen. 2007. "A History of Behavior Genetics." In *Neurobehavioral Genetics*, edited by B. C. Jones and P. Mormède, 1–16. Boca Raton, FL: CRC Press.

McClearn, G. E. 1970. "Behavioral Genetics." *Annual Review of Genetics* 4: 437–68.

McClearn, G. E., and W. Meredith. 1966. "Behavioral Genetics." *Annual Review of Psychology* 17: 515–50.

McClearn, Gerald E., and John C. DeFries. 1973. *Introduction to Behavioral Genetics.* San Francisco: W. H. Freeman.

McGue, Matt. 2008. "The End of Behavioral Genetics?" *Acta Psychologica Sinica* 40: 1073–87.

McGue, Matt, Thomas J. Bouchard Jr., William G. Iacono, and David T. Lykken. 1993. "Behavioral Genetics of Cognitive Ability: A Life-Span Perspective." In *Nature, Nurture, and Psychology*, edited by R. P. G. E. McClearn, 59–76. Washington, DC: American Psychological Association.

McGue, Matt, and David T. Lykken. 1992. "Genetic Influence on Risk of Divorce." *Psychological Science* 3: 368–73.

McGuire, T. R. 2008. "Jerry Hirsch (1922–2008)—Trailblazer and Teacher." *Genes, Brain, and Behavior* 7: 834–35.

Medvetz, Thomas. 2012. *Think Tanks in America*. Chicago: University of Chicago Press.

Mehler, Barry. 1997. "Beyondism: Raymond B. Cattell and the New Eugenics." *Genetica* 99: 153–63.

Mehler, Barry, and Keith Hurt. 1998. "Race Science and the Pioneer Fund." http://www.ferris.edu/ISAR/Institut/pioneer/search.htm.

Mekel-Bobrov, Nitzan, Sandra L. Gilbert, Patrick D. Evans, Eric J. Vallender, Jeffrey R. Anderson, Richard R. Hudson, Sarah A. Tishkoff, and Bruce T. Lahn. 2005. "Ongoing Adaptive Evolution of ASPM, a Brain Size Determinant in *Homo sapiens*." *Science* 309: 1720–22.

Mello, Felicia. 2007. "Financial Choices Influenced by Genes." *Boston Globe*, October 8.

Mensh, Elaine, and Harry Mensh. 1991. *The IQ Mythology: Class, Race, Gender, and Inequality*. Carbondale: Southern Illinois University Press.

Merton, Robert King. 1973. *The Sociology of Science: Theoretical and Empirical Investigations*. Chicago: University of Chicago Press.

Mialet, Helene. 2003. "Review: The 'Righteous Wrath' of Pierre Bourdieu." *Social Studies of Science* 33: 613–21.

Miele, Frank. 2002. *Intelligence, Race, and Genetics: Conversations with Arthur R. Jensen*. Boulder, CO: Westview.

———. 2008. "En-Twinned Lives." *Skeptic* 14: 27–37.

Miller, Adam. 1994. "The Pioneer Fund: Bankrolling the Professors of Hate." *Journal of Blacks in Higher Education*: 58–61.

———. 1995. "Professors of Hate." In *The Bell Curve Debate: History, Documents, Opinions*, edited by R. Jacoby and N. Glauberman, 162–78. New York: Times Books.

Miller, David. 1995. "Introducing the 'Gay Gene': Media and Scientific Representations." *Public Understanding of Science* 4: 269–84.

Mirowski, Philip. 2011. *Science-Mart: Privatizing American Science*. Cambridge, MA: Harvard University Press.

Mirowski, Philip, and Robert Van Horn. 2005. "The Contract Research Organization and the Commercialization of Scientific Research." *Social Studies of Science* 35: 503–48.

Mitroff, Ian I. 1974. "Norms and Counter-Norms in a Select Group of the Apollo Moon Scientists: A Case Study of the Ambivalence of Scientists." *American Sociological Review* 39: 579–95.

Mody, Cyrus. 2011. *Instrumental Community: Probe Microscopy and the Path to Nanotechnology*. Cambridge, MA: MIT Press.

Montagu, Ashley. 1964. *The Concept of Race*. New York: Collier Books.

Moore, David S. 2001. *The Dependent Gene*. New York: W. H. Freeman.

Moore, Kelly. 2008. *Disrupting Science*. Princeton, NJ: Princeton University Press.

Morton, Newton. E. 1972. "Human Behavioral Genetics." In *Genetics, Environment, and Behavior*, edited by L. Ehrman, G. S. Omenn, and E. Caspari. New York: Academic Press.

Muir, Hazel. 2001. "Divorce Is Written in the DNA." *New Scientist*, July.

Mulkay, Michael J. 1976. "Norms and Ideology in Science." *Social Science Information* 15: 637–56.

Munsinger, Harry. 1977. "A Reply to Kamin." *Behavior Genetics* 7: 407–9.

Mustanski, B. S., M. G. Dupree, C. M. Nievergelt, S. Bocklandt, N. J. Schork, and D. H. Hamer. 2005. "A Genomewide Scan of Male Sexual Orientation." *Human Genetics* 116: 272–78.

Naik, Gautam. 2011. "Mistakes in Scientific Studies Surge." *Wall Street Journal*, August 10.

———. 2013. "A Genetic Code for Genius?" *Wall Street Journal*, February 15, C3.

National Institutes of Mental Health. 2006. *The Numbers Count: Mental Disorders in America*. Washington, DC: National Institutes of Mental Health.

Naureckas, Jim. 1995. "Racism Resurgent: How Media Let the Bell Curve's Pseudo-Science Define the Agenda on Race." *Extra!*, January/February.

Neale, M. C., and L. R. Cardon. 1992. *Methodology for Genetic Studies of Twins and Families*. Dordrecht: Kluwer.

Neisser, U., G. Boudoo, T. J. Bouchard, A. W. Boykin, N. Brody, S. J. Ceci, D. F. Halpern, J. C. Loehlin, R. Perloff, R. J. Sternberg, and S. Urbina. 1996. "Intelligence: Knowns and Unknowns." *American Psychologist* 51: 77–101.

Nelkin, Dorothy. 1984. *The Creation Controversy: Science or Scripture in the Schools*. Boston: Beacon Press.

———. 1987. *Selling Science: How the Press Covers Science and Technology*. New York: W. H. Freeman.

———. 1999. "Behavioral Genetics and Dismantling the Welfare State." In *Behavior Genetics: The Clash of Culture and Biology*, edited by R. A. Carson and M. A. Rothstein, 156–71. Baltimore: Johns Hopkins University Press.

Nelkin, Dorothy, and M. Susan Lindee. 1995. *The DNA Mystique: The Gene as a Cultural Icon*. New York: W. H. Freeman.

———. 2000. *The DNA Mystique: The Gene as a Cultural Icon*. Ann Arbor: University of Michigan Press.

Nelson, Nicole. 2011. "Capturing Complexity: Experimental Systems and Epistemic Scaffolds in Animal Behavior Genetics." Thesis, Science and Technology Studies, Cornell University, Ithaca, NY.

Nestle, Marion. 2002. *Food Politics: How the Food Industry Influences Nutrition and Health*. Berkeley: University of California Press.

News Editorial Staffs. 2003. "The Runners-Up." *Science* 302: 2039–45.

Newton-Cheh, C. and J. N. Hirschhorn. 2005. "Genetic Association Studies of Complex Traits: Design and Analysis Issues." *Mutation Research* 573: 54–69.

Nichols, P. L., and V. Elving Anderson. 1973. "Intellectual Performance, Race, and Socioeconomic Status." *Social Biology* 20: 367–74.

Nisbett, Richard E. 2010. *Intelligence and How to Get It.* New York: Norton.

Nuffield Council on Bioethics. 2002. "Genetics and Human Behaviour: The Ethical Context." Nuffield Council on Bioethics, London.

Nyborg, H. 2003. "The Sociology of Psychometric and Bio-Behavioral Sciences: A Case Study of Destructive Social Reductionism and Collective Fraud in 20th Century Academia." In *The Scientific Study of General Intelligence: Tribute to Arthur R. Jensen*, edited by H. Nyborg, 441–502. New York: Pergamon.

O'Connell, A. N. 2001. *Models of Achievement: Reflections of Eminent Women in Psychology.* Vol. 3. London: Lawrence Erlbaum Associates.

Oettl, Alexander. 2012. "Sociology: Honour the Helpful." *Nature* 489: 496–97.

Orbach, Susan, Joe Schwartz, and Mike Schwartz. 1974. "The Case for Zero Heritability." *Science for the People* 6: 23–25.

Oreskes, Naomi. 2004. "The Scientific Consensus on Climate Change." *Science* 306: 1686.

Oreskes, Naomi, and Erik M. Conway. 2010. *Merchants of Doubt: How a Handful of Scientists Obscured the Truth on Issues from Tobacco Smoke to Global Warming.* New York: Bloomsbury Press.

Osborne, R. H., and B. T. Osborne. 1999a. "The Founding of the Behavior Genetics Association, 1966–1971." *Social Biology* 46: 207–18.

———. 1999b. "The History of the Journal *Social Biology*, 1954–1999." *Social Biology* 46: 164–93.

———. 1999c. "Interdisciplinary Symposia Published in *Social Biology*." *Social Biology* 46: 232–64.

Page, E. B. 1972. "Behavior and Heredity." *American Psychologist* 27: 660–61.

Panofsky, Aaron. Forthcoming. "Rethinking Scientific Authority: Race Controversies in Behavior Genetics." In *Creating Authority*, edited by C. Calhoun and R. Sennett. New York: New York University Press.

Panofsky, Aaron L. 2009. "Behavior Genetics and the Prospect of 'Personalized Social Policy.'" *Policy and Society* 28: 327–40.

———. 2011. "Field Analysis and Interdisciplinary Science: Scientific Capital Exchange in Behavior Genetics." *Minerva* 49: 295–316.

Parens, Erik. 2004. "Genetic Differences and Human Identities: On Why Talking about Behavioral Genetics Is Important and Difficult." Hastings Center, Garrison, NY.

Parens, Erik, Audrey Chapman, and Nancy Press. 2006. *Wrestling with Behavioral Genetics.* Baltimore: Johns Hopkins University Press.

Parsons, P. A. 1967. *The Genetic Analysis of Behavior*. London: Methuen.

Pastore, Nicholas. 1949. *The Nature-Nurture Controversy*. New York: Kings Crown Press.

Paul, Diane B. 1987. "The Nature-Nurture Controversy: Buried Alive." *Science for the People* 19: 17–20.

———. 1994. "Dobzhansky in the 'Nature-Nurture' Debate." In *The Evolution of Theodosius Dobzhansky: Essays on His Life and Thought in Russia and America*, edited by M. B. Adams, 219–31. Princeton, NJ: Princeton University Press.

———. 1998a. *Controlling Human Heredity: 1865 to the Present*. Amherst, NY: Humanities Books.

———. 1998b. "Eugenic Origins of Medical Genetics." In *The Politics of Heredity: Essays on Eugenics, Biomedicine, and the Nature-Nurture Debate*, edited by Diane B. Paul, 133–56. Albany: State University of New York Press.

———. 1998c. *The Politics of Heredity: Essays on Eugenics, Biomedicine, and the Nature-Nurture Debate*. Albany: State University of New York Press.

———. 1998d. "The Rockefeller Foundation and the Origins of Behavior Genetics." In *The Politics of Heredity: Essays on Eugenics, Biomedicine, and the Nature-Nurture Debate*, edited by Diane B. Paul, 53–80. Albany: State University of New York Press.

Pinker, Steven. 2002. *The Blank Slate*. New York: Viking.

Plomin, R., R. Corley, J. C. DeFries, and D. W. Fulker. 1990. "Individual Differences in Television Viewing in Early Childhood: Nature as Well as Nurture." *Psychological Science* 1: 371–77.

Plomin, R., and D. Daniels. 1987. "Why Are Children from the Same Family So Different from One Another?" *Behavioral and Brain Sciences* 10: 1–16.

Plomin, Robert. 1979. "Training in Behavioral Genetics: A Survey of BGA Members." *Behavior Genetics* 9: 419–24.

———. 1990. "The Role of Inheritance in Behavior." *Science* 248: 183–88.

Plomin, Robert, J. C. DeFries, and J. C. Loehlin. 1977. "Genotype-Environment Interaction and Correlation in the Analysis of Human Behavior." *Psychological Bulletin* 84: 309–22.

Plomin, Robert, J. C. DeFries, and Gerald McClearn. 1980. *Behavioral Genetics, a Primer*. New York: W. H. Freeman.

Plomin, Robert, and John C. DeFries. 1985. *Origins of Individual Differences in Infancy*. New York: Academic Press.

Plomin, Robert, John. C. DeFries, Ian W. Craig, and Peter McGuffin. 2003a. "Behavioral Genetics." In *Behavioral Genetics in the Postgenomic Era*, edited by R. Plomin, J. C. DeFries, I. W. Craig, and P. McGuffin, 3–15. Washington, DC: American Psychological Association.

———. 2003b. *Behavioral Genetics in the Postgenomic Era*. Washington, DC: American Psychological Association.

Plomin, Robert, John. C. DeFries, Valerie S. Knopik, and Jenae M. Neiderhiser. 2013. *Behavioral Genetics*. New York: Worth Publishers.

Plomin, Robert, John. C. DeFries, Gerald E. McClearn, and Michael Rutter. 2008. *Behavioral Genetics*. New York: Worth Publishers.

Plomin, Robert, Michael J. Owen, and Peter McGuffin. 1994. "The Genetic Basis of Complex Human Behaviors." *Science* 264: 1733–39.

Pollan, Michael. 2009. *In Defense of Food: An Eater's Manifesto*. New York: Penguin.

Popp Berman, Elizabeth. 2012. *Creating the Market University*. Princeton, NJ: Princeton University Press.

Popper, Karl. 1962. *Conjectures and Refutations*. New York: Basic Books.

Proctor, Robert, and Londa Schiebinger, eds. 2008. *Agnotology: The Making and Unmaking of Ignorance*. Stanford, CA: Stanford University Press.

Provine, William. 1986. "Geneticists and Race." *American Zoologist* 26: 857–87.

Quinn, William G., William A. Harris, and Seymour Benzer. 1974. "Conditioned Behavior in *Drosophila melanogaster.*" *Proceedings of the National Academy of Sciences of the United States of America* 71: 708–12.

Rabinow, Paul, and Gaymon Bennett. 2012. *Designing Human Practices*. Chicago: University of Chicago Press.

Rafter, Nicole Hahn. 1997. *Creating Born Criminals*. Urbana: University of Illinois Press.

Ramsden, Edmund. 2002. "Carving Up Population Science: Eugenics, Demography, and the Controversy over the 'Biological Law' of Population Growth." *Social Studies of Science* 32: 857–99.

———. 2008. "Eugenics from the New Deal to the Great Society: Genetics, Demography, and Population Quality." *Studies in History and Philosophy of Science Part C: Studies in History and Philosophy of Biological and Biomedical Sciences* 39: 391–406.

Reardon, Jenny. 2005. *Race to the Finish*. Princeton, NJ: Princeton University Press.

Revkin, Andrew C. 2012. "Peter Gleick Admits to Deception in Obtaining Heartland Climate Files." *Dot Earth, New York Times*. http://dotearth.blogs .nytimes.com/2012/02/20/peter-gleick-admits-to-deception-in-obtaining-heartland-climate-files/, accessed August 15, 2013.

Rhea, Sally-Ann, and Robin P. Corley. 1994. "Applied Issues." In *Nature and Nurture during Middle Childhood*, edited by J. C. DeFries, R. Plomin, and D. W. Fulker, 295–309. Cambridge, MA: Blackwell.

Rheinberger, Hans-Jörg. 1997. *Toward a History of Epistemic Things*. Stanford, CA: Stanford University Press.

Rice, G., C. Anderson, N. Risch, and G. Ebers. 1999. "Male Homosexuality: Absence of Linkage to Microsatellite Markers at Xq28." *Science* 284: 665–67.

Richardson, Sarah S. 2011. "Race and IQ in the Postgenomic Age: The Microcephaly Case." *BioSocieties* 6: 420–46.

Ridley, Matt. 2003. *Nature Via Nurture.* New York: HarperCollins.

Risch, Neil, Richard Herrell, Thomas Lehner, Kung-Yee Liang, Lindon Eaves, Josephine Hoh, Andrea Griem, Maria Kovacs, Jurg Ott, and Kathleen Ries Merikangas. 2009. "Interaction between the Serotonin Transporter Gene (5-Httlpr), Stressful Life Events, and Risk of Depression: A Meta-Analysis." *JAMA: The Journal of the American Medical Association* 301: 2462–71.

Risch, Neil J., and David Botstein. 1996. "A Manic Depressive History." *Nature Genetics* 12: 351–53.

Robertson, Miranda. 1989. "False Start on Manic Depression." *Nature* 342: 222.

Rogaev, Evgeny Ivanovich. 2012. "Genomics of Behavioral Diseases." *Frontiers in Genetics* 3: 1–4.

Rose, Nikolas. 1992. "Engineering the Human Soul: Analyzing Psychological Expertise." *Science in Context* 5: 351–69.

———. 2000. "The Biology of Culpability: Pathological Identity and Crime Control in a Biological Culture." *Theoretical Criminology* 4: 5–34.

Rose, Richard J. 1995. "Genes and Human Behavior." *Annual Review of Psychology* 46: 625–54.

Rose, Steven P. R. 1995. "The Rise of Neurogenetic Determinism." *Nature* 373: 380–82.

———. 1998. "Neurogenetic Determinism and the New Euphenics." *British Medical Journal* 317: 1707–8.

Rose, Steven P. R., J. Hambley, and J. Harwood. 1973. "Science, Racism, and Ideology." *Socialist Register.*

Rosenfield, Patricia L. 1992. "The Potential of Transdisciplinary Research for Sustaining and Extending Linkages between the Health and Social Sciences." *Social Science and Medicine* 35: 1343–57.

Rosenthal, David. 1968. "The Genetics of Intelligence and Personality." In *Biology and Behavior: Genetics,* edited by D. G. Glass, 69–78. New York: Rockefeller University Press.

Roubertoux, Pierre L. 2008. "Jerry Hirsch (20 September 1922–3 May 2008): A Tribute." *Behavior Genetics* 38: 561–64.

Rowe, David C. 1994. *The Limits of Family Influence: Genes, Experience, and Behavior.* New York: Guilford.

———. 1997. "A Place at the Policy Table? Behavior Genetics and Estimates of Family Environmental Effects on IQ." *Intelligence* 24: 133–58.

———. 2002. *Biology and Crime.* Los Angeles: Roxbury.

Royce, Joseph R., and Leendert P. Mos. 1979. *Theoretical Advances in Behavior Genetics.* Alphen aan den Rijn, The Netherlands: Sijthoff and Noordhoff.

Rudwick, Martin. 1985. *The Great Devonian Controversy: The Shaping of Sci-*

entific Knowledge among Gentlemanly Specialists. Chicago: University of Chicago Press.

Rushton, J. Philippe. 1994. *Race, Evolution, and Behavior: A Life History Perspective.* New Brunswick, NJ: Transaction Publishers.

———. 1998. "The New Enemies of Evolutionary Science." *Liberty* 2: 31–35.

———. 2000. *Race, Evolution, and Behavior: A Life History Perspective.* Port Huron, MI: Charles Darwin Research Institute.

Rushton, J. Philippe, David W. Fulker, Michael C. Neale, David K. B. Nias, and Hans J. Eysenck. 1986. "Altruism and Aggression: The Heritability of Individual Differences." *Journal of Personality and Social Psychology* 50: 1192–98.

Rushton, J. Philippe, and Arthur R. Jensen. 2005. "Thirty Years of Research on Race Differences in Cognitive Ability." *Psychology, Public Policy, and Law* 11: 235–94.

Rutter, Michael. 2002. "Nature, Nurture, and Development: From Evangelism through Science toward Policy and Practice." *Child Development* 73: 1–21.

———. 2006. *Genes and Behavior.* Malden, MA: Blackwell.

Rutter, Michael, and Robert Plomin. 1997. "Opportunities for Psychiatry from Genetic Findings." *British Journal of Psychiatry* 171: 209–19.

Sandel, Michael J. 2007. *The Case against Perfection.* Cambridge, MA: Harvard University Press.

Scarr, Sandra. 1981. *Race, Social Class, and Individual Differences in IQ.* Hillsdale, NJ: Lawrence Erlbaum Associates.

———. 1987. "Three Cheers for Behavior Genetics: Winning the War and Losing Our Identity." *Behavior Genetics* 17: 219–28.

———. 1992. "Developmental Theories for the 1990s: Development and Individual Differences." *Child Development* 63: 1–19.

———. 1993. "Biological and Cultural Diversity: The Legacy of Darwin for Development." *Child Development* 64: 1333–53.

Scarr, Sandra, and Louise Carter-Saltzman. 1979. "Twin Method: Defense of a Critical Assumption." *Behavior Genetics* 9: 527–42.

Scarr, Sandra, A. J. Pakstis, S. H. Katz, and W. B. Barker. 1977. "Absence of a Relationship between Degree of White Ancestry and Intellectual Skills within a Black Population." *Human Genetics* 39: 69–86.

Scarr, Sandra, and R. A. Weinberg. 1976. "IQ Test Performance of Black Children Adopted by White Families." *American Psychologist* 31.

Scarr-Salapatek, Sandra. 1976. "Review of Kamin's 'The Science and Politics of IQ.'" *Contemporary Psychology* 21: 98–99.

Schiff, Michel, and Richard C. Lewontin. 1986. *Education and Class: The Irrelevance of IQ Genetic Studies.* Oxford: Oxford University Press.

Schönemann, Peter H. 2001. "Better Never Than Late: Peer Review and the Preservation of Prejudice." *Ethical Human Sciences and Services* 3: 7–21.

Schwartz, M., and J. Schwartz. 1974. "Evidence against a Genetical Component to Performance on IQ Tests." *Nature* 248.

Schwartz, Michael, and Joseph Schwartz. 1976. "Comment on 'IQs of Identical Twins Reared Apart.'" *Behavior Genetics* 6: 367–68.

Scott, John Paul. 1947. "The Conference on Genetics and Social Behavior at the Roscoe B. Jackson Memorial Laboratory." *American Psychologist* 2: 176–77.

———. 1976. "The Future of Behavior Genetics." In *Human Behavior Genetics*, edited by A. R. Kaplan, 33–48. Springfield, IL: Charles C. Thomas.

———. 1985. "Investigative Behavior: Toward a Science of Sociality." In *Leaders in the Study of Animal Behavior*, edited by D. A. Dewsbury, 389–429. London: Associated University Presses.

Scott, John Paul, and John L. Fuller. 1965. *Genetics and the Social Behavior of the Dog*. Chicago: University of Chicago Press.

Scrinis, Gyorgy. 2008. "On the Ideology of Nutritionism." *Gastronomica* 8: 39–48.

Scull, Andrew. 1989. *Social Order/Mental Disorder: Anglo-American Psychiatry in Historical Perspective*. Berkeley: University of California Press.

Sedgwick, J. 1995. "Inside the Pioneer Fund." In *The Bell Curve Debate*, edited by R. Jacoby and N. Glauberman, 144–61. New York: Times Books.

Segal, Nancy L. 1993. "Twin, Sibling, and Adoption Methods: Tests of Evolutionary Hypotheses." *American Psychologist* 48: 943–56.

———. 2012. *Born Together—Reared Apart: The Landmark Minnesota Twin Study*. Cambridge, MA: Harvard University Press.

Segerstråle, Ullica Christina Olofsdotter. 2000. *Defenders of the Truth: The Battle for Science in the Sociobiology Debate and Beyond*. Oxford; New York: Oxford University Press.

Seigel, Judy. 1999. "Expert: Kennedys Have Risk-Taking Gene." *Jerusalem Post*, July 19, 5.

Seligman, Daniel 1994. "A Substantial Inheritance." *National Review*, October 10, 56–60.

Sennett, Richard. 1980. *Authority*. New York: Knopf.

Shapin, Steven. 1994. *A Social History of Truth: Civility and Science in Seventeenth-Century England*. Chicago: University of Chicago Press.

———. 2008. *The Scientific Life*. Chicago: University of Chicago Press.

Shapin, Steven, and Simon Schaffer. 1985. *Leviathan and the Air-Pump: Hobbes, Boyle, and the Experimental Life: Including a Translation of Thomas Hobbes, Dialogus Physicus De Natura Aeris by Simon Schaffer*. Princeton, NJ: Princeton University Press.

Shenk, David. 2010. *The Genius in All of Us*. New York: Doubleday.

Sherman, Stephanie L., John C. DeFries, Irving I. Gottesman, John C. Loehlin, Joanne M. Meyer, Mary Z. Pelias, John Rice, and Irwin Waldman. 1997. "Behavioral Genetics '97: ASHG Statement, Recent Developments in Human

Behavioral Genetics: Past Accomplishments and Future Directions." *American Journal of Human Genetics* 60: 1266–75.

Sherwood, J. J., and M. Nataupsky. 1968. "Predicting the Conclusions of Negro-White Intelligence Research from Biographical Characteristics of the Investigators." *Journal of Personality and Social Psychology* 8: 53–58.

Shockley, William B. 1969. "A Proposal That the U.S. Government Query the National Academy of Sciences Regarding Possible Updating of the Academy's 1967 Statement on Human Genetics and Urban Slums." *US Congressional Record* 115: 40528–29.

———. 1992. *Shockley on Eugenics and Race*. Edited by R. Pearson. Washington, DC: Scott-Townsend.

———. 1992 [1968]. "Proposed Research to Reduce Racial Aspects of the Environment-Heredity Uncertainty." In *Shockley on Eugenics and Race*, edited by R. Pearson, 94–104. Washington, DC: Scott-Townsend.

———. 1992 [1974]. "Eugenic or Anti-Dysgenic, Thinking Exercises." In *Shockley on Eugenics and Race*, edited by R. Pearson, 130–67. Washington, DC: Scott-Townsend.

Shorter, Edward. 1997. *A History of Psychiatry*. New York: John Wiley and Sons.

Shostak, Sara. 2013. *Exposed Science: Genes, the Environment, and the Politics of Population Health*. Berkeley: University of California Press.

Shostak, Sara, Jeremy Freese, Bruce G. Link, and Jo C. Phelan. 2009. "The Politics of the Gene: Social Status and Beliefs about Genetics for Individual Outcomes." *Social Psychology Quarterly* 72: 79–93.

Shurkin, Joel N. 2008. *Broken Genius: The Rise and Fall of William Shockley, Creator of the Electronic Age*. New York: Palgrave Macmillan.

Simon, Bart. 2002. *Undead Science: Science Studies and the Afterlife of Cold Fusion*. New Brunswick, NJ: Rutgers University Press.

Siontis, Konstantinos C. M., Nikolaos A. Patsopoulos, and John P. A. Ioannidis. 2010. "Replication of Past Candidate Loci for Common Diseases and Phenotypes in 100 Genome-Wide Association Studies." *European Journal of Human Genetics* 18: 832–37.

Sismondo, Sergio. 2009. "Ghosts in the Machine: Publication Planning in the Medical Sciences." *Social Studies of Science* 39: 171–98.

———. 2011. "Bourdieu's Rationalist Science of Science: Some Promises and Limitations." *Cultural Sociology* 5: 83–97.

Skinner, B. F. 1953. *Science and Human Behavior*. New York: Free Press.

Snyderman, Mark, and Stanley Rothman. 1988. *The IQ Controversy, the Media, and Public Policy*. New Brunswick, NJ: Transaction Publishers.

Society for Neuroscience. 2003. "Guidelines for Crisis Management." Society for Neuroscience, Washington, DC.

Spear, Joseph Howard, Jr. 2000. "The Sociology of Reductionism: A Case Study of the 'Neuroscience Explosion.'" PhD diss., Sociology, University of Virginia, Richmond.

Spuhler, J. N. 1967. *Genetic Diversity and Human Behavior.* Chicago: Aldine.

Spuhler, J. N., and G. Lindzey. 1967. "Racial Differences in Behavior." In *Behavior-Genetic Analysis*, edited by J. Hirsch. New York: McGraw-Hill.

Stampnitzky, Lisa. 2013. *Disciplining Terror: How Experts Invented "Terrorism."* Cambridge: Cambridge University Press.

Starr, Paul. 1982. *The Social Transformation of American Medicine.* New York: Basic Books.

Steele, Claude M., and Joshua Aronson. 1995. "Stereotype Threat and the Intellectual Test Performance of African Americans." *Journal of Personality and Social Psychology* 69: 797–811.

Sternberg, R. J. 1985. "The Black-White Differences and Spearman's *g*: Old Wine in New Bottles That Still Doesn't Taste Good." *Brain and Behavioral Sciences* 8: 244.

Stoltenberg, Scott F., and Margit Burmeister. 2000. "Recent Progress in Psychiatric Genetics—Some Hope but No Hype." *Human Molecular Genetics* 9: 927–35.

Stoolmiller, Mike. 1999. "Implications of the Restricted Range of Family Environments for Estimates of Heritability and Nonshared Environment in Behavior—Genetic Adoption Studies." *Psychological Bulletin* 125: 392–409.

Sullivan, P. 2011. "Don't Give Up on GWAS." *Molecular Psychiatry*: 1–2.

Sullivan, P. F. 2007. "Spurious Genetic Associations." *Biological Psychiatry* 61: 1121–26.

Tabery, James. 2009. "Making Sense of the Nature-Nurture Debate." *Biology and Philosophy* 24: 711–23.

Taubes, Gary. 2007. *Good Calories, Bad Calories.* New York: Vintage.

Taylor, Charles. 1985. *Philosophical Papers.* Vol. 1, *Human Agency and Language.* Cambridge: Cambridge University Press.

———. 1993. "To Follow a Rule . . ." In *Bourdieu: Critical Perspectives*, edited by C. Calhoun, E. LiPuma, and M. Postone, 45–60. Chicago: University of Chicago Press.

Taylor, H. F. 1980. *The IQ Game.* New Brunswick, NJ: Rutgers University Press.

Tesser, A. 1993. "The Importance of Heritability in Psychological Research: The Case of Attitudes." *Psychological Review* 100: 129–42.

Thoday, J. M., and A. S. Parkes. 1968. *Genetic and Environmental Influences on Behavior.* Edinburgh: Oliver and Boyd.

Thomas, W. I., and Dorothy S. Thomas. 1928. *The Child in America: Behavior Problems and Programs.* New York: Knopf.

Tizard, J. 1976. "Progress and Degeneration in the 'IQ Debate': Comments on Urbach." *British Journal for the Philosophy of Science* 27: 251–58.

Truett, K. R., L. J. Eaves, E. E. Walters, A. C. Heath, J. K. Hewitt, J. M. Meyer,

J. Silberg, M. C. Neale, N. G. Martin, and K. S. Kendler. 1994. "A Model System for Analysis of Family Resemblance in Extended Kinships of Twins." *Behavior Genetics* 24: 35–49.

Tucker, William H. 1994. *The Science and Politics of Racial Research*. Urbana: University of Illinois Press.

———. 2001. "Bankrolling Racism: 'Science' and the Pioneer Fund." *Race and Society* 4: 195–205.

———. 2002. *The Funding of Scientific Racism: Wickliff Draper and the Pioneer Fund*. Urbana: University of Illinois Press.

Turkheimer, E. 2011. "Commentary: Variation and Causation in the Environment and Genome." *International Journal of Epidemiology* 40: 598–601.

Turkheimer, Eric. 2000. "Three Laws of Behavioral Genetics and What They Mean." *Current Directions in Psychological Science* 9: 160–64.

———. 2004. "Spinach and Ice Cream: Why Social Science Is So Difficult." In *Behavior Genetics Principles: Perspectives in Developments, Personality, and Psychopathology*, edited by L. DiLalla, 161–89. Washington, DC: American Psychological Association.

———. 2006. "Mobiles: A Gloomy View of the Future of Research into Complex Human Traits." In *Wrestling with Behavioral Genetics*, edited by E. Parens, A. Chapman, and N. Press, 165–78. Baltimore: Johns Hopkins University Press.

Turkheimer, Eric, Andreana Haley, Mary Waldron, Brian D'Onofrio, and Irving I. Gottesman. 2003. "Socioeconomic Status Modifies Heritability of IQ in Young Children." *Psychological Science* 14: 623–25.

Turner, Stephen. 2000. "Disciplinarity and Its Other." In *Practising Interdisciplinarity*, edited by P. Weingart and N. Stehr, 46–65. Toronto: University of Toronto Press.

UNESCO. 1969. *Four Statements on the Race Question*. Paris: UNESCO.

Urbach, Peter. 1974a. "Progress and Degeneration in the 'IQ Debate' (I)." *British Journal for the Philosophy of Science* 25: 99–135.

———. 1974b. "Progress and Degeneration in the 'IQ Debate' (II)." *British Journal for the Philosophy of Science* 25: 235–59.

van den Oord, Edwin J. C. G., and Patrick F. Sullivan. 2003. "False Discoveries and Models for Gene Discovery." *Trends in Genetics* 19: 537–42.

Vandenberg, S. G., and J. C. DeFries. 1970. "Our Hopes for Behavior Genetics." *Behavior Genetics* 1: 1–2.

Vandenberg, Steven G. 1965. *Methods and Goals in Human Behavior Genetics*. New York: Academic Press.

Vernon, Philip E., John R. Dill, Irving I. Gottesman, Irving Biederman, and Kenneth E. Clark. 1970. "Testing Negro Intelligence: Comments on Eysenck." *The Humanist*: 34–37.

Vetta, Atam. 1977. "Genetical Concepts and IQ." *Social Biology* 24: 166–68.

Wachs, Theodore D. 1983. "The Use and Abuse of Environment in Behavior-Genetic Research." *Child Development* 54: 396–407.

Wachs, Theodore D., and Robert Plomin. 1991. "Conceptualization and Measurement of Organism-Environment Interaction." Washington, DC: American Psychological Association.

Wade, Nicholas. 2009. "The Evolution of the God Gene." *New York Times*, November 14.

Wahlsten, Douglas. 1990. "Insensitivity of the Analysis of Variance to Heredity-Environment Interaction (with Open Peer Commentary)." *Behavioral and Brain Sciences* 13: 109–61.

———. 1995. "Review: J. P. Rushton, *Race, Evolution, and Behavior.*" *Canadian Journal of Sociology* 20: 129–33.

———. 2003. "Airbrushing Heritability." *Genes, Brain, and Behavior* 2: 1–3.

———. 2008. "Obituary: Jerry Hirsch (1922–2008)." *Genes, Brain, and Behavior* 7: 833–34.

Wasserman, David. 1996. "Symposium: Genetics and Crime, University of Maryland Conference." *Politics and the Life Sciences* 15: 107–9.

Watson, James D. 2003. "A Molecular Genetics Perspective." In *Behavioral Genetics in the Postgenomic Era*, edited by R. Plomin, J. C. DeFries, I. W. Craig, and P. McGuffin, xxi–xxii. Washington, DC: American Psychological Association.

Watson, John B. 1930. *Behaviorism*. Chicago: University of Chicago Press.

Weber, Max. 1946. "Science as a Vocation." In *From Max Weber*, edited by H. H. Gerth and C. W. Mills, 122–56. New York: Oxford University Press.

Weiner, Jonathan. 1999. *Time, Love, Memory*. New York: Alfred A. Knopf.

Weisz, G., A. Cambrosio, P. Keating, T. Schlich, L. Knaapen, and V. Tournay. 2007. "The Emergence of Clinical Guidelines." *Milbank Quarterly* 75: 691–727.

Whitley, Richard. 2000. *The Intellectual and Social Organization of the Sciences*. Oxford: Oxford University Press.

Whitney, Glayde. 1995a. "Ideology and Censorship in Behavior Genetics." *Mankind Quarterly* 35: 327–43.

———. 1995b. "Presidential Address to the Behavior Genetics Association: Twenty-Five Years of Behavior Genetics." *Mankind Quarterly* 35: 327–42.

———. 2000. "Reply to Winston and Peters' 'On the Presentation and Interpretation of International Homicide Data . . .'" *Psychological Reports* 86: 1234–36.

Wilayto, Phil. 1997. "The Bell Curve: Roadmap to the 'Ideal' Society." http://www.mediatransparency.org/story.php?storyID=8. Vol. 2005, Media Transparency.

Williams, Trevor. 1977. "Reply to Professor Lewontin's Letter." *Behavior Genetics* 7: 37.

Wilson, Edward O. 1975. *Sociobiology: The New Synthesis.* Cambridge, MA: Belknap Press of Harvard University Press.

———. 1978. "The Attempt to Suppress Human Behavioral Genetics." *Journal of General Education* 29: 277–87.

Wilson, James Q., and Richard J. Herrnstein. 1985. *Crime and Human Nature.* New York: Simon and Schuster.

Winston, Andrew, and Michael Peters. 2000. "On the Presentation and Interpretation of International Homicide Data." *Psychological Reports* 86: 865–71.

Wright, Susan. 1986. "Molecular Biology or Molecular Politics? The Production of Scientific Consensus on the Hazards of Recombinant DNA Technology." *Social Studies of Science* 16: 593–620.

Wright, William. 1998. *Born That Way: Genes, Behavior, Personality.* New York: Alfred A. Knopf.

Yong, Ed. 2013. "Chinese Project Probes the Genetics of Genius." *Nature* 497: 297–99.

Zammit, S., N. Wiles, and G. Lewis. 2010a. "The Study of Gene-Environment Interactions in Psychiatry: Limited Gains at Substantial Cost?" *Psychological Medicine* 40: 711–16.

Zammit, Stanley, Michael J. Owen, and Glyn Lewis. 2010b. "Misconceptions about Gene-Environment Interactions in Psychiatry." *Evidence Based Mental Health* 13: 65–68.

Ziman, J. M. 2000. *Real Science: What It Is, and What It Means.* Cambridge; New York: Cambridge University Press.

Zimmer, Carl. 2012. "A Sharp Rise in Retractions Prompts Calls for Reform." *New York Times*, April 16.

Index